RISING TO THE CHALLENGE
of the National Science Education Standards
The Processes of Science Inquiry
Primary Grades

Dr. Karen Ostlund
Science Education Center
The University of Texas at Arlington
Arlington, Texas

Sheryl Mercier
Kings Canyon Unified School District
Reedley, California

ISBN 0-9658768-1-0

S & K Associates
Squaw Valley, California
email: sKmer5@aol.com

RISING TO THE CHALLENGE
of the National Science Education Standards
The Processes of Science Inquiry
Primary Grades

This book was set in Tekton by S & K Associates.
The authors and editors were Karen Ostlund and Sheryl Mercier.
The illustrations were done by Sheryl Mercier.
Printed in the United States of America.

First Printing 1999
Second Printing 2001
Third Printing 2003
Fourth Printing 2004

TABLE OF CONTENTS

Introduction to the Processes of Inquiry

Seeds - Connect to the Standards   38

INVITATION TO INQUIRY

If science is presented as a collection of facts, concepts, and laws, students are left with the perception that everything that there is to be known about science has already been discovered and can be found in their textbook. This misrepresents the nature of science. Science is constantly changing. As research is conducted and advanced technology enhances the ability to make observations, our understanding of the world is modified. An awareness of the fluid nature of science helps students to recognize that change is the rule rather than the exception. Science as a process of inquiry guides students to comprehend the dynamic nature of science.

Hands-on and minds-on experiences are fundamental to the inquiry approach to science. Teaching science as inquiry involves a shift from depending on the textbook as the basic source of information to using the textbook as a reference. Laboratory activities are central as students investigate and inquire about the world. Their own observations become the authoritative source of data. Students discover the facts, concepts, and laws of science in much the same way as the original discoveries were made. The emphasis on firsthand observations in learning science reflects the belief that students should model the processes used by scientists in learning facts and concepts.

This discovery approach places greater emphasis on the logical thinking processes by which new knowledge is acquired and less emphasis on the rote learning of information. With the rapid rate at which knowledge is expanding, it is becoming difficult to prescribe exactly which facts and concepts should be transmitted to elementary students. It makes more sense to teach students the processes of science to enable them to construct their own knowledge. These processes prepare students to acquire new knowledge.

The idea of discovery is central to science as inquiry. The nature of science itself is a process of discovery. Science as inquiry requires that the learning environment be set up in which students can engage in hands-on and minds-on science discovery activities. Finally, science as inquiry involves students in using science processes to investigate and discover patterns in the world.

IMPLEMENTING THE NATIONAL SCIENCE STANDARDS

Implementing the National Science Education Standards involves providing opportunities for students to construct their own understandings about science. Students should be encouraged to ask questions about nature and seek answers, collect things, count and measure objects, make qualitative observations, organize collections and observations, and discuss findings. These experiences will bring out the idea of consistency in nature. Students should be permitted to repeat observations and investigations to compare what happens under different circumstances. These activities will stimulate curiosity and encourage students to develop an interest in their environment and how nature works.

As students continue to investigate their world, explaining inconsistency can strengthen the idea of consistency. When students observe differences in the way things work or get different results in repeated investigations, it should be pointed out that different findings could lead to interesting new questions that can be investigated.

In addition to understanding the dynamic nature of science, students need to develop attitudes that characterize the enterprise of science such as curiosity, willingness to suspend judgment, open-mindedness, and skepticism. Curiosity is the inspiration for doing science. Scientists are question-askers; any observation they make may be the catalyst for an investigation. Scientists develop hypotheses and draw inferences from direct observation of natural phenomena and withhold judgments until all the data have been accumulated. They must be open-minded enough to set aside previously held views in light of new information. Given the dynamic nature of science, scientists must be willing to challenge accepted dogmas. They must exhibit a healthy skepticism toward any conclusion that is not supported by careful observations. By setting the stage for students to develop these attitudes, you are facilitating their acquisition of knowledge and skills in science as well as implementing the vision of the National Science Education Standards.

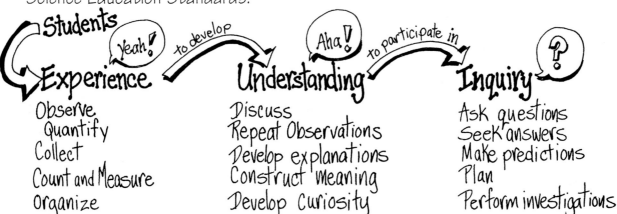

Students — (Yeah!) to develop — (Aha!) to participate in — (?)

Experience	Understanding	Inquiry
Observe	Discuss	Ask questions
Quantify	Repeat Observations	Seek answers
Collect	Develop explanations	Make predictions
Count and Measure	Construct meaning	Plan
Organize	Develop Curiosity	Perform investigations

THE PROCESS SKILLS: THE KEY TO SUCCESS IN SCIENCE

The processes of science that students use to carry out hands-on and minds-on investigations and experiments are the focus of science as inquiry. Brief descriptions of these processes are:

Observing - the use of the five senses to gather data about objects and events

Communicating -the use of the spoken and written words, graphs, drawings, and diagrams to share information and ideas with others

Comparing - the use of observations to ascertain similarities and differences in objects and events

Classifying -grouping objects or events according to similar properties

Measuring -the use of standard or nonstandard units to determine length, mass, volume, time, etc.

Predicting - the use of data to forecast future events based on observations and inferences

Inferring - the use of a logical thought process to show a relationship between observations or provide an explanation of an observation

Defining Operationally - a definition framed in terms of your experiences

Formulating Models - developing a conceptual or physical representation of an object or event

Investigating

Formulating Hypotheses - making an educated guess about the relationship of manipulated and responding variable that can be tested experimentally

Controlling Variables - identifying and controlling variable in order to determine their effect on the outcome of an experiment

Experimenting - hypothesizing, designing an experiment to test the hypothesis, controlling variables, interpreting the data collected, and drawing conclusions

Interpreting Data - analyzing and synthesizing data in order to draw a conclusion

Relating - the use of a logical thought process to determine the relationships involving interactions, dependencies, and cause-and-effect between and among objects and events

Applying -the use of a logical thought process to put scientific knowledge to use

3

4 Introduction to the Processes

OBSERVATION
The Basic Process Skill of Science

Careful observation is a critically important step in scientific investigations from the primary grades to the most advanced levels. In these activities, primary students will be observing similarities and differences in various objects. They will also be observing changes in an assortment of objects. Observing activities are based on familiar content - shape, color, number, position, order, etc.

The process of observing is basic to the development of all the other inquiry skills. Observations are fundamental to any scientific investigation; these observations lead to the construction of inferences or hypotheses that can be tested by observational techniques. Therefore, observing provides both a basis for new inferences and hypotheses and a tool for testing existing inferences and hypotheses.

The process of observing involves using several or all of the senses - sight, touch, hearing, taste, and smell. Observing requires that you pick up objects, feel them, shake them, press them, and do all the things which help you obtain information about the object. It does not mean visual inspection alone.

Quantitative observations (those that use numbers) usually communicate more information than qualitative observations (those that use words). Exact measurements are quantitative observations; however, estimations or comparisons also clarify observations.

When making observations, it is important to notice the things that are changing. Compare properties before, during, and after the change. Observe the rate of change and the circumstances under which the change is observed. Observations are properties that are directly perceived by the senses; an inference involves an interpretation of direct observations.

5

COOPERATIVE/COLLABORATIVE LEARNING GROUPS

Cooperative learning involves students working together in small groups to help one another master skills and subject matter content. Research indicates that using cooperative group instructional strategies increases student achievement and social skill acquisition (Johnson, 1986). Developing social skills is an important prerequisite for academic learning since achievement will improve as students become more effective in working with each other. It is within cooperative group situations where there is academic material that must be mastered that social skills become most relevant. One of the most important advantages of cooperative learning tasks is that social skills are required, used, reinforced, and mastered.

We are not born with instinctive behaviors that help us interact in socially acceptable ways the first time we are placed in a situation in which we make contact with others. If we expect students to work together, we must teach them how to use social skills when working together just as purposefully and precisely as academic skills are taught. The acquisition of social skills is essential for building and maintaining an enduring family, a successful career, and lasting friendships. Therefore, learning in the context of cooperative groups requires students to develop and use the social skills that will be necessary for leading fruitful and fulfilling lives as adults.

How can student social skill acquisition be assessed in the context of cooperative learning? The purpose of assessment has important implications for what information is collected and how it is gathered. Assessment involves many different ways of gathering information about students in order to plan instruction at a challenging but developmentally appropriate level for pupils. __What__ are the social skills that should be assessed? __How__ should these social skills be assessed? These are the questions this section attempts to answer.

__What__ social skills should be assessed in cooperative learning groups? The social skills listed below are divided into three categories: (1) **cluster skills** which help students form a group; (2) **task skills** which help the group complete the task; and (3) **camaraderie skills** which help group members like each other when the task is finished.

The **cluster skills** facilitate moving into groups quickly and quietly as well as beginning the group activity. The **task skills** have a content focus and help the group meet its subject matter objective as well as helping the group work effectively to create a high quality product. **Task skills** begin with skills that help groups manage completing the task, through skills that maximize the mastery and retention of the material, to skills that build deeper level understanding through the use of critical thinking skills. The **camaraderie skills** help students feel better about themselves, each other, and the group to build group cohesiveness and stability.

Cluster Skills	*Helps students form a group*

- Move into groups quietly.
- Stay with your group.
- Use quiet voices.
- Address people in your group by name.
- Look at the speaker.
- Listen actively.
- Look at the group's paper or project.
- Keep hands and feet to self.
- Share materials.
- Take turns.
- Encourage everyone to participate.
- Contribute ideas.

Task Skills	*Helps group complete the task*

- Ask questions.
- Ask for help.
- Ask for clarification.
- Offer to explain or clarify.
- Check for understanding.
- Review the instructions.
- State and/or restate the purpose of the assignment.
- Set or call attention to time limits.
- Offer procedures on how to most effectively do the task.
- Paraphrase other members' contributions.
- Clarify other members' contributions.
- Allow each person to speak once before speaking again.
- Summarize the material aloud.
- Pursue accuracy by correcting and/or adding to summaries.
- Seek elaboration by relating to other learning or knowledge.
- Search for clever ways of remembering ideas and facts.
- Encourage vocalization of other members' reasoning processes.
- Ask other members to plan out loud how to solve a problem or make a decision.
- Differentiate when there is disagreement.
- Integrate different ideas into a single position.
- Listen to all ideas before reaching consensus.
- Ask for justification of others' conclusions or ideas.
- Extend other members' answers or conclusions.
- Probe by asking in-depth questions that lead to deeper analysis.
- Generate further answers.
- Test reality by checking the group's work against instructions.

Camaraderie Skills	*Helps students like each other when task is complete*

- Avoid put-downs.
- Give each member of your group a compliment.
- Express support and acceptance verbally and/or nonverbally.
- Express warmth and liking (toward group members and group).
- Describe feelings when appropriate.
- Energize the group with humor, ideas, or enthusiasm.
- Relieve tension by joking.
- Criticize ideas without criticizing people.

It is best to stress only <u>one</u> social skill each time a group engages in a cooperative group activity. The social skill should be determined by observing the operation of the groups to pick out behaviors which are not constructive to effective functioning by the group as a whole or by individual group members. Once a skill has been identified which is appropriate for the cooperative group activity and the social development of the students, the teacher can explain, model, and elicit other examples of appropriate behaviors for the specific skill. For example, when teaching students to use quiet voices, the teacher could ask the students to place several fingers on their neck over the vocal chords, say something out loud to feel the vibration produced by the vocal chords, and then whisper something to feel that there is almost no vibration in the vocal chords.

When teaching st udents to encourage everyone to participate, each group member could be given counting chips of a different color which would be placed in the center of the group's work space each time someone encouraged another group member to participate. The different colored chips could be counted at the end of the activity to determine how many times each group member engaged in this behavior. The counting chips could be used again to teach students to allow each person to speak once before speaking again. This time each group member would receive one counter that would be used to take turns talking. When someone talks, s/he would place a counter in the middle of the group's work space and would not speak again until all four counters are in the center at which point the counters would be distributed back to each group member to used when talking again.

Establishing the goals for the cooperative classroom is a prerequisite to setting up cooperative learning activities. The expectations must be made clear and explicit so that complex tasks, such as using higher level collaborative skills become automatic. These three expectations are effective in establishing the goals in a cooperative classroom:

> Help me teach.
> Help others learn.
> Be prepared to learn.

The goals for the cooperative classroom determine the classroom rules that need to be observed in order to reach those goals. Classroom rules should be limited and stated in a positive manner. Establishing the norm behaviors expected in every cooperative learning activity is the first step in setting up assessment of the social skill behaviors. The following norm behaviors are effective in establishing the protocol for a cooperative classroom:

> Listen.
> Be responsible for yourself.
> Respect others.
> Stay on task.

Inappropriate behaviors that are not conducive to group functioning can be addressed by referring to these expectations. The first expectation, <u>Listen</u>, stops any talking when the teacher or any other person in the group is talking because one cannot listen and talk at the same time. The second expectation encourages each student to take responsibility for his/her behavior such as following directions, bringing required materials to class, and being seated and prepared to learn when the class begins. The third expectation takes care of one student interfering with another student with physical and/or verbal contact. Finally, the last expectation addresses any off task behavior such as talking about what's for lunch or engaging in manipulating materials in an inappropriate manner.

How should social skills be assessed? Since the purpose of assessment has important implications for what information is collected and how it is gathered, assessment can involve many different ways of gathering information about students. One method of assessing students working in cooperative groups is to provide each group with a "ruler" such as the one shown below:

+5	+4	+3	+2	+1	0	-1	-2	-3	-4	-5
Names _____						_____				

Cooperative group members place this "ruler" in the center of their workspace and put their names on it. This method of assessment awards five bonus points to each group engaging in the norm behaviors listed above and practicing the designated social skill for the particular cooperative group activity. The teacher monitors students' behavior when the cooperative groups start working. When a group is not engaging in the appropriate social skills, the teacher explains which behavior the students need to exhibit and demonstrates how to perform the social skill. The teacher crosses out the **+5** on the "ruler" so the group now has four bonus points for the activity. This method of assessing social skills shifts the responsibility for engaging in appropriate behavior from the teacher to the group members. The bonus points may be used in a variety of ways to reward each group for appropriate social skill behaviors: extra credit on the cooperative group assignment, viewing a special science video, participating in a demonstration for the class, etc.

Some observations that have been made about using this method for assessing social skill development include the observance that a teacher seldom has to cross out more than one or two numbers on the "ruler." If students notice that the teacher is monitoring their group, they encourage each other to exhibit appropriate social skills. In some situations <u>one</u> particular student in a cooperative group may lack basic social skills. After several attempts are made by the teacher to encourage the other members of the group to provide constructive support for the student to engage in appropriate behaviors, it may be necessary to give that particular student his/her own individual "ruler" while the rest of the group continues to use a group "ruler." Generally, the use of the "ruler" is not an ongoing process of assessment throughout the entire school year. It is effective at the beginning of the year for getting started with assessing social skill behaviors in cooperative groups. However, after four to six weeks, students are usually exhibiting effective working relationships and it might not be necessary to use the instrument to award points/scores to each group. Social skill development sometimes retrogrades in the middle of the school year; so teachers might want to use the "rulers" for a few weeks to help remind students about expectations for behaviors.

One way to ensure interdependence is to assign complementary and interconnected roles to group members. We recommend having students work in pairs at the primary level. When materials/supplies are limited, you could put two pairs together to form groups of four. However, primary students often haven't developed the social skills necessary to work with more than one partner. Each partner is assigned a responsibility that the group needs to function. These include a **Communications Manager** to manage the learning task and make sure that his/her partner understands what is being learned, and to record and report information gathered to the other pairs and the teacher; and a **Materials Manager** to distribute the needed materials for the group, and to keep track of time. The job descriptions are listed below.

Communications Manager

The Communications Manager is in charge of reading or repeating the instructions, checking the activity results, asking informational questions of the teacher, and conducting discussions about processes and results. Note: Only the "CM" from each pair is allowed to ask the teacher informational questions after the pair have discussed the assignment and has been unable to answer a question together. This limits the number of students asking questions about the assignment so that a teacher will have questions from only 10-12 students rather than 20-24 students. The "CM" either conducts the activity or assigns his/her partner to carry out the activity.

The Communications Manager is also in charge of collecting the information and recording it on the class table, graph, or other data collection instrument used in the activity. The "CM" in cooperation with his/her partner is also responsible for certifying the results that are recorded by the pair on the data sheets. If partners gather data individually, the "CM" collects the individual data sheets and hands them in to the teacher for the pair. The "CM" is responsible for reporting the results of the group activity to the class. This can be an oral report given from his/her seat or a written report placed on a class summary chart on the chalkboard. If his/her partner has any special concerns or comments about the data collected, the "CM" is also responsible for giving this information before the discussion of results begins.

Materials Manager

The Materials Manager is responsible for collecting and returning all materials and equipment for the science activity, assembling and operating equipment, and checking the activity results. The "MM" is the only student who will be out of his/her seat during the activity. After the lesson is introduced, directions are given, and questions are answered, the "MM" gathers the materials and returns to the group workstation to set up the equipment for the activity. When group clean up is completed; the "MM" returns the materials and equipment to the supply area. The Materials Manager is also in charge of keeping track of time, watching for group safety, encouraging his/her partner, monitoring the noise level, and checking the activity results.

One of the partners can even be assigned to monitor specific behaviors. For example, a partner may tally how many times each member of the person speaks in order to determine if everyone in the group is contributing ideas. **The Materials Manager** may be assigned the additional responsibility of noise monitor for the group by giving group members a signal when a quiet voice is not used. Student observers can be used to get even more extensive data on each group's functioning. However, try not to count too many different behaviors at one time. At first just keep track of who talks in each pair to get a participation pattern for the groups. It is also a good idea for the teacher to collect notes on specific student behaviors so that the frequency data is extended. When it is obvious that group members lack certain collaborative skills they need in order to cooperate with each, the teacher will want to intervene in order to help the members learn these collaborative skills. However, teachers should not intervene any more than is absolutely necessary because pairs can often work their way through their own problems (task and social skill) and acquire not only a solution, but also a method of solving similar problems in the future. The best time to teach cooperative skills is when the students need them. It is important that the cooperative skills be taught in the context of the class where they are going to be used, or are practiced in that setting, because transfer of skill learning from one situation to another cannot be assumed.

Product

The product required from the group may be a report, a single set of answers that all members of the group agree to, the average of individual paper or test scores, or the number of group members reaching specific criteria. Whatever the measure, the learning of group members needs to be evaluated by a criteria-referenced system. Besides assessing students on how well they learned the assigned concepts and information, group members should also receive feedback on how effectively they collaborated. Two grades may be given - one for achievement and one for collaborative behavior. Collaborative skills should focus both on members' contributions to each other's learning and to the maintenance of effective working relationships among group members.

Closure

Closure to the lesson can be provided by summarizing the major points in the lesson, asking students to recall ideas or give examples, or answering any final questions students have. At the end of the lesson, students should be able to summarize what they have learned and to understand where they will use it in future lessons. Besides assessing students on how well they learned the assigned material, group members should also receive feedback on how effectively they worked together. Two grades could be given for the activity: one for achievement and one for how well the group functioned (social skill behaviors). Social skills should focus both on members' contributions to each other's learning and to the maintenance of effective working relationships among group members. Provide closure to the cooperative learning activity by asking groups to state one social skill they performed well as they worked together and one social skill they need to work on the next time the group meets.

Using cooperative learning groups in science is an effective method of utilizing people and materials efficiently. In a class ranging in size from 24 to 30 students, only one-fourth of the materials you would normally need are required to involve students in hands-on science. Research indicates that students working cooperatively learn interpersonal skills, improve personal responsibility and learn concepts as well as or better than if they had worked on the science activity individually.

The overall goal of social skill acquisition is positive, on-task students who enjoy their time together, care about each other, and produce high quality work.

Communication Manager

1. I read instructions.
2. I ask the teacher questions if we do not understand.
3. I check and report results.
4. I record information.
5. I work well with others.

fold

Materials Manager

1. I collect and return materials.
2. I assemble and operate the equipment.
3. I check the results.
4. I check time and safety.
5. I work well with others.

 Learning Groups

INTRODUCING THE PROCESSES OF INQUIRY
Overhead Transparencies- Teaching Tips

The one-page activity sheets that follow are designed as overhead transparencies for your use in directing students' introduction to the inquiry skills of science. Use the activities to experience and discuss each individual skill as a guided whole group lesson. Allow time for metacognition (thinking about their thinking).

Observe With Our Eyes

Purpose: The purpose of developing children's skill of observation is so that they will be able to use all their senses (appropriately and safely) to gather relevant information for their investigations of things around them. This will guide primary students to begin to distinguish the relevant from the irrelevant in the context of a particular investigation or problem. If children are not able to make this distinction, they may miss significant information and narrow the focus of their observation too soon. Encourage students to make as many observations as they can, giving attention to detail and not just gross features. Young children are able to do very well this in relation to objects that interest and intrigue them.

Materials: color crayons for each student, butcher or chart paper, a moveable toy (a toy that moves by batteries or mechanical mechanisms or a toy that you can roll, slide, etc.)

Teaching Tips: Have students observe the toy and use colors to draw a picture of it. Encourage students to observe and draw every part of the toy. Then draw a large outline of the toy on butcher paper. Invite students one at a time to add one part to the picture until the large drawing accurately shows every detail of the toy. As each student draws a part, ask the other students to describe the part. Use descriptions to generate a word list on the chalkboard or chart paper. After completing the group drawing, have students use words from the word list the write sentences describing the toy.

Encourage students to look carefully and try to notice everything. Then move the toy (or wait for it to move) to get a different view. Ask students to describe what they see using drawings and/or words. Ask students to compare what they see with what you drew or wrote.

Metacognition: Ask, "What is the most unique thing you noticed when you looked at your object or event?"

Observe with Our Ears

Purpose: Children can't see sounds - they have to associate the sounds they hear with the object producing the sound. Children need to make connections between sounds and sources. This activity develops children's ability to identify sounds and discriminate between sounds. Matching different sounds is valuable as a pre-reading experience, teaching a child that words s/he cannot see have referents in reality.

Materials: opaque plastic containers with tops or lids (film canisters, margarine tubs, plastic eggs, etc.), a variety of sound sources, i.e., gravel, pebbles, beads, seeds, marbles, pennies, sand, rice, paper clips, etc.)

Teaching Tips: Fill sets of two containers with the same kind of object(s). Leave enough room in each container so that the object(s) can move and produce sound. Have students shake the containers to match the two containers that sound alike. Ask students to describe the sounds and generate a word list on the chalkboard or chart paper.

  Ask students to listen carefully and try to find out what is making the sound. Explain that this is the <u>sound source</u>. Ask them to listen to the sound more than one time (they are the <u>sound receivers</u>). Then have students compare the sound to other sounds they have heard. Ask students to draw what they think made the sound.

Metacognition (Thinking about your Thinking): Ask, "How can you find out what is making a sound?"

Observe With Our Nose

Purpose: To combine sensory awareness and language development

Materials: several different essence oils (available from craft/hobby stores for potpourri) or extracts, construction paper, marker for drawing a circle in the upper left hand corner of each sheet of construction paper

Teaching Tips: The sense of smell can play a part in space orientation. You can find food by following your nose. Use a medicine dropper to place several drops of one of the extracts in a circle on each sheet of white construction paper. Each student will receive one of the scented papers. Demonstrate how to waft in order to smell an unknown substance. Advise students that the smells on the paper are not harmful. Then give students permission to smell the circle and describe the smell. Use student descriptions to generate a word list on the chalkboard or chart paper. Then have students draw something that might have those smell and use words from the word list to describe the smell.

Emphasize to students that they should never hold an unknown substance under their nose to smell it without permission! Demonstrate how to hold a substance away from your nose and use your hand to wave over the substance toward your nose. (This is called wafting.) Give students permission to hold the paper under their nose and take a shallow breath.
Tell students that if the substance does not irritate them, they may hold the substance under their nose and take a deep breath. Have students describe the smell.

Metacognition (Thinking about Your Thinking):
Ask students, "How can smells help you?" (Warn about food that is bad or dangerous situations such as fire, etc.)

Observe With Our Tongue

Purpose: To combine sensory awareness and language development

Materials: fresh fruits and vegetables (apples, oranges, potatoes, grapes, lemons, carrots, cucumber, peas, tomatoes, etc.)

Teaching Tips: Cut a hole in each side of a brown grocery bag large enough for children to put their hands in but not see the objects. Students feel and describe the form and texture of the fruits and vegetables. Children then associate the way fruits and vegetables feel on the outside with the way they taste on the inside. After the feel-bag activity, cut the fruits and vegetables into chunks and let the children taste them. Ask students: How does the vegetable feel in your mouth? What sounds go along with eating it? Describe the vegetable's taste. Generate a word list using students' responses on the chalkboard or chart paper (sweet, sour, bitter, salty, soft, hard, crunchy, chewy, mushy, etc.). Have students draw a picture of a fruit or vegetable and use words from word list to describe the texture, form, taste, and sound.

Tell students that they should never taste anything without permission! Remind them to make sure your hands and the food is clean. Instruct them to place the food in their mouth but not to chew it yet. Tell students to use their tongue to explore the food. Have them chew the food (if necessary).
Then ask students to describe the taste of the food.

Metacognition: Ask, "What can you find out by tasting?"

Purpose: To develop observational abilities and language precision. Matching different textures is valuable as a pre-reading experience, teaching a child that words s/he cannot see have referents in reality.

Materials: socks, common objects such as Ping-Pong ball, half-cup measure, toy car, plastic spoon, plastic cube, etc.

Teaching Tips: Have students feel classroom objects that are rough, smooth, hard, and soft (chalkboard, wall, rug, etc.). Ask students to describe the textures and shapes of the objects in order to generate a word list on a chalkboard or chart paper. Initially, place one familiar object inside each sock. After children develop the skill to describe and name objects, place three or four objects in each sock. Have children feel the object(s) in the sock, describe what they feel to a friend or an adult, and if possible, name the objects. The child can check a guess immediately by looking in the sock. Later, try objects of similar shapes but different textures, perhaps an apple, a tennis ball, and an orange. In all feeling-sock activities, emphasize describing the object as a means of encouraging language skills; naming the object is secondary. You may also want to have students draw a picture of the object and write words to describe the texture and form from the word list generated.

Tell students not to peek in the sock (it will spoil the fun)! Instruct them to carefully touch the object with their fingertips. Have them rub the object in different places. Tell student to press on the object. Suggest that they squeeze the object between your fingers. Have students describe how the object feels.

Metacognition (Thinking about your Thinking):
Ask, "What can touching tell you about an object?"

Communicate

Purpose: Just as important as careful observation is the ability to communicate. Verbal communication forms the vehicle for conducting most of the classroom activities. In some activities, precise verbal communication is the objective. In other activities, including those in observing, different objectives may be stressed, but communication still forms the vehicle by which exchange of information occurs. Precise language is needed, whether you are describing an observation reporting a measurement, interpreting data, or performing other of the science processes of inquiry. As you teach the process skills, you will observe improvement in your students' verbal communication skills in science and this improvement should carry over to other parts of the curriculum.

Materials: Plan Ahead! Have each student bring in a teddy bear (have some extra bears for students who do not bring one in for the activity).

Teaching Tips: Ask each student to describe his/her teddy bear to a partner. Tell students to get ready to do the "Teddy Bear March." Have one student describe the attributes of his/her teddy bear, i.e., my teddy bear is brown, has a <u>blue</u> shirt, red pants, etc. The next student has to find one attribute that is like an attribute of the first teddy bear, i.e.; my teddy bear has a <u>blue</u> pair of pants. Then have students tell a partner how they are like their teddy bear and how they are different from their teddy bear. Record the class data on a chart and make a graph to show the results. Ask students to write a statement to tell what the graph is showing.

Metacognition: What are all the ways that you communicated about your bear?

Classify

Purpose: Young children are born collectors. Rocks of every size, shape, and description are often collected because they can be found almost everywhere. As children look at, describe, and compare their rocks, these experiences provide built-in language development. Children can devise their own classification schemes. The younger ones will notice color and size first. Rocks can also be classified by how they feel to the touch. Have children describe the feel of a rock (slippery, smooth, bumpy, rough, sharp, pebbly, etc.). The general shape of rocks is another means of classification. Some rocks are more round than others, some are jagged and angular, and some are flat. Manipulating the rocks can bring about an awareness of the hardness or softness of rocks. The rocks can be rubbed against each other or scratched with a fingernail, nail, or a penny in order to compare and classify them. The rocks can also be compared wet or dry. The children can take the rocks outside to see which ones can be used to draw streaks or lines on the (dry) sidewalk. If the rocks are soft enough, the children may be able to make distinctive colored streaks with them and use this as another classification scheme. The rocks can also be held in the sunshine to compare the shine or luster.

Materials: a collection of different kinds of rocks

Teaching Tips: Have students observe the properties of the rocks carefully. Then have them find out how the rocks are alike and how the rocks are different. Ask students to choose one property of the rocks.
Then have them divide the rocks into two groups: the rocks that have that property and the rocks that don't have that property. Remind students to label each group.

Metacognition: Ask, "How could you divide your objects into two groups using another property?"

Measure

Purpose: When young children ask the questions "How long is it?" or "How big is it?" they provide teachers with an opportunity to introduce measurement at a level the child will understand. In order to measure, children need to be able to divide the unit to be measured into subunits of similar length; and to substitute one part of the measuring unit on the object being measured the appropriate number of times. The ability to measure follows the understanding of number as a concept. To understand number, children should have experiences in classifying, comparing, and ordering. Children should be familiar with the terminology of *as many as, more than, the same as,* and *as long as.* Not only must they understand, but they must also be comfortable with the use of the words. Children need to work in the area of nonstandard measurement when they are very young. The answer to the question "How long is it?" requires an answer saying "As long as. . ." There always has to be something to compare it with.

Teaching Tips:
Show children how you can determine the length of a table by measuring it with your outstretched palm. Ask children for ideas for using their bodies to measure the width of the room. They could use the length of their bodies, the number of strides in walking, or the number of their own foot lengths putting one in front of the other. Allow many opportunities for children to measure distances using their

bodies. Discuss the idea that a measurement unit can be anything you want it to be as long as you *name* the unit and measure only with that unit.

Invite students to find an object in the classroom to measure. Then have students select something (a unit) to use to measure the object. Ask them to guess (estimate) how many units the object is. Ask students to compare their objects to their units. Then have them count how many units the object is. Finally, have students compare their measurements with their guesses (estimates).

Metacognition: How accurate was your estimate? How could you measure your object using a different unit?

Infer

Purpose: Use observations as the basis for making inferences

Materials: approximately 4 inch squares of fabric with different textures, approximately 4 inch squares of paper, color crayons

Teaching Tips:
Give children textured fabric samples and ask them to observe them using their eyes and their sense of touch. Have the children make rubbings of each fabric sample by placing a piece of paper over the fabric and gently rubbing the paper with a color crayon. Then have the students infer which fabric was used to make each rubbing. Ask students to carefully observe the fabrics and the rubbings. Ask them to match each fabric with its rubbing and explain the observations that helped them identify the fabric that was used to make each rubbing. Then have students make another rubbing to see if it matches the rubbing that they matched to the fabric. Allow students to repeat the process until they feel confident about their inferences.

Metacognition: What are some other explanations for your observations?

Predict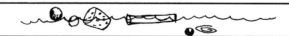

Purpose: Sort out what effect size and shape have on floating

Materials: plastic dishpan, 2 sets of floating and non-floating objects - corks, marbles, etc. in a Zip-lock® bag for each pair of students

Teaching Tips: Ask students to observe and sort the set of objects into two groups: objects that they think will **float** in water and objects that they think will **sink** in water. Have students place the first set of objects in a plastic dishpan of water to discover whether they sink or float. Show students how to place the objects on a laminated chart.

Then have students observe the objects in the second set and predict whether the second set of similar objects will sink or float based on their observations of the first set of objects. Have students test their predictions by placing the objects in the water to see if they sink or float. Explain to students that we can think about how to use information we gather to make more accurate predictions. Ask students what information they can use to make more accurate predictions.

Metacognition: How could you modify your prediction?

Define Operationally

Purpose: Children will define terms in the context of their own experiences. We ask children to use their own observations, thoughts, and experiences to find out about the world around them. However, teachers have vastly more experience and background to draw upon, so there is a pitfall that is ever present as you teach the skill of defining operationally. The pitfall rests in deciding when the children's definitions are adequate. A definition, which seems adequate to a child, at least according to his/her experiences and observations, may not be adequate to the teacher. The child is not wrong from his/her point of view. To correct the child by requiring him/her to use your definition, which is more adequate, from your point of view, but not based on the child's experience, would be to defeat a major aspect of this process skill.

As a child matures, s/he will meet progressively more complex terms which need to be defined. S/he will also need to revise terms previously defined to improve the definitions according to new experiences. To a child, an operational definition can only be stated in terms of activities the child has done using words, ideas, skills, and operations which are mastered - a doing definition. The skill is to be developed gradually through the child's continuing experience as a higher level of competence is acquired in the science process of inquiry. When the definition is inadequate, let the child discover this inadequacy as s/he tries to use the definition.

Teaching Tips: Ask students what the word "fastener" means. Have the children look at the clothes they are wearing. Are there any fasteners? Write their responses on the chalkboard or chart paper so you can read back what they say. Have the children show each fastener they discover and talk about it. (Buttons, zippers, belt buckles, shoe laces, Velcro closures, shoe buckles, bobby pins, barrettes, hair ribbons, safety pins, clasps on jewelry) Ask questions such as "What keeps the hair from Susan's eyes?" or "How does Sarah keep the bracelet on her arm?" List all the fasteners that students have found. Then have students draw pictures of different fasteners and write their own operational definition of a fastener.

Metacognition: How could you revise your definition?

Make a Model

Purpose: To use the body in creative drama and models

Materials: pictures of animals or objects from magazines or books

Teaching Tips: Point out pictures of different animals or objects and ask students to demonstrate how the animal or object moves. Have a student lead the class and suggest ways to move. Have the class follow behind in single file, moving as the leader suggests. Then have the leader fall to the end of the line and the next child becomes the leader.

Suggestions for movements: waddle like a duck, fly like a jet, hop like a rabbit or squirrel, slither like a snake, sail like a kite, float like a seed or soap bubble, fly like a bee or butterfly, soar like a hawk or crow, walk like an ostrich, hop like a kangaroo, stalk like a lion, crawl like a seal or walrus, move like a dog or cat. Explain that they are modeling how an animal or object moves and that a model represents an actual object. Prompt students to think of other models of actual objects. You may want to list these on the chalkboard. Ask students how they are alike and how they are different from the animal or object that they are modeling.

Metacognition: How are the actual thing and the model alike? Different?

Investigate

Purpose: To focus on the processes of doing investigations to develop the ability to ask scientific questions, investigate aspects of the world around us, and use observations to construct reasonable explanations for the questions posed

Materials: ice cubes, plastic containers, insulating materials

Teaching Tips: Start by asking lots of questions and listening intently to the answers, encouraging the children to hypothesize, guess, ask questions of you. Go slowly pacing to the children's interest level. Piaget tells us that young children learn by doing, thinking about what they have done, and discussing this with others. This verbal interaction is essential; it is through the exchange of ideas that the child organizes and reorganizes his/her thinking, and learns.

How long will an ice cube last if it is put on a plate?
Will an ice cube melt faster if it is broken up?
Will an ice cube that is twice as large take twice as long to melt?
Will an ice cube last longer in a glass of water at room temperature or in the room temperature air?
Does the size of the glass of water make any difference in the time it takes the ice cube to melt?
Does the temperature of a glass of water make any difference in the time it takes the ice cube to melt?
Can you design a container that will keep an ice cube from melting for a long time?
What materials could you use for insulation to keep the ice cube from melting?
Who can keep an ice cube from melting? (without putting it in the freezer)
Would your container also work well for keeping hot things from cooling down?
How could you design an investigation to answer that question?

Metacognition: How accurate was your guess? Could you do the investigation another way? What other questions would you like to answer? Was this a fair test? What could you do to make this a fair test?

Observe with our Eyes

1. Look with your eyes.

2. Watch the object as it moves.

3. Tell what you see. Write and draw.

4. Look again. Compare what you see with what you wrote and drew.

Observe with our Ears

1. Listen carefully to the sound.
2. Compare the sound to sounds you have heard.
3. Tell about the sound.
4. Draw what you think made the sound.

Observe with our Nose

1. Ask before you smell.
2. Hold the smell away from your face.
3. Push the smell to your nose.
4. Smell and compare.
5. Tell about the smell.

Observe with our Tongue

1. Ask before you taste.
2. Wash your hands.
3. Put the food on your tongue.
4. Taste then chew.
5. Tell about the taste.

         Introduction to the Processes

Observe with our Skin

1. Touch carefully with your fingers.
2. Rub the object in different places.
3. Squeeze the object.
4. Tell how the object feels.

Communicate

Bears
squishy
furry
large small
soft
fuzzy
light

Alike	Different
4 legs	big little
2 eyes	colors
2 ears	heavy light
1 nose	eye color
4 paws	sit or stand
stuffed	

My Bear

has 2 black eyes
4 legs, pink
tongue

8
7
6
5
4
3
2
1
0
Black White Brown Pink

1. Observe an object or event.
2. Tell about what you observe.
   look, touch, smell, taste, hear
3. Make a record by drawing, writing
   charting, graphing, counting.

Classify

1. Observe the objects.

2. Find out how they are alike and how they are different.

3. Choose one thing about them.

4. Put the objects into groups.

Measure

1. Find an object to measure.
2. Pick something to use for a unit.
3. Guess how many units the object is.
4. Compare the object to your unit.
5. Count how many units the object is.
6. Compare the measurement with your guess.

Infer

1. Observe an object or something happening.
2. Think about what you observed.
3. Explain your observations.
4. Make more observations to compare.
5. Think and explain again.

     Introduction to the Processes

Predict

1. Observe and think about what you know.
2. Predict. Tell about what you think will happen.
3. Test your prediction. Make observations
4. Compare your observations.
5. Make a more accurate prediction.

Define Operationally

What is a fastener?

1. Observe an object or event.

2. Think about your observations.

3. Describe the object or event by telling what you observed.

Introduction to the Processes

Make a Model

1. Observe something.
2. Think how you could show it.
3. Make your model.
4. Compare the model to the real thing.

Investigate

1. Ask a question.
2. Guess the answer.
3. Set up materials to find the answer to your question.
4. Test and observe carefully to find out what happens.
5. Tell what happened.

Science and Reading/Language Skills

Science can be a vehicle for developing language. Students need to be able to make observations in science and communicate those observations. Vocabulary can be introduced informally into discussion by using the context of the discussion to provide the meaning for the word. For example, students may notice that there is a baby plant inside a seed providing the opportunity to introduce the precise word *embryo*. Vocabulary can be introduced formally as students discover different kinds of seeds and want to know the name of each seed type. Word walls can be used to help students with reading and language skills. A piece of chart paper, the chalkboard, or butcher paper can be used to display descriptive words elicited from students during discussions. If you put an icon next to the word as you write it on the word wall, it will help students learn the word. Post the word wall where all students have access to it so that they can use words from the list on their activity sheets.

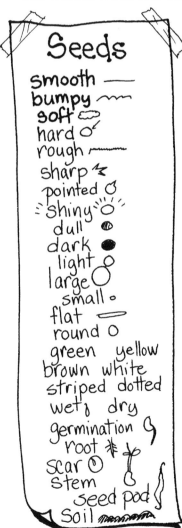

Seeds

smooth
bumpy
soft
hard
rough
sharp
pointed
shiny
dull
dark
light
large
small
flat
round
green   yellow
brown white
striped dotted
wet   dry
germination
root
scar
stem
seed pod
soil

Matter

red orange yellow
green blue purple
brown black pink
odor
smell
odorless
sweet
sour
bitter
salty
hard
soft
square
rectangle
round
flat
large   small
striped   spots
sticky   swirls

Fossils

dinosaurs
tall I long
armored
skin
tail
jaw   teeth
spikes
horns
clubs
claws
skeleton
fossil
footprints
eggs
nest
plant eater
meat eater
paleontologist

37

SEEDS

One or two days before you plan to begin, ask students to gather and bring in small objects that they think might be seeds. Students may bring in packages of seeds sold for gardens; apple seeds, peach pits, nuts, and seeds from trees, bushes, and weeds. Collect some additional objects, most of which are seeds but a few that are not seeds such as cinnamon candy drops, gravel, and cereal. Place one of each kind of "seed" in a snack-sized zip-lock® bag for each student. Have them add their "seeds" to the collection and compare the seeds in the bag with the seed they brought. Encourage students to ask questions and devise ways of finding their own answers; such as, "What's inside a seed? Does a seed have a top and a bottom? How can you be sure this is a seed?" It is a good sign of learning when students say, "Let's try it. Let's plant it. Let's grow it." When students seek the test of direct information, they are engaging at their own level in the nature of science - they are developing a belief in their own ability to answer questions.

National Science Education Standards

Unifying Concepts and Processes: Constancy, Change, and Measurement	
Characteristics of Organisms	Organisms have basic needs; plants require air, water, nutrients, and light. Plants have different structures that serve different functions in growth, survival, and reproduction.
Abilities Necessary to do Scientific Inquiry	As students focus on the processes of inquiry, they develop the ability to ask scientific questions, investigate aspects of the world around them, and use their observations to construct reasonable explanations for the questions posed.
Understanding about Scientific Inquiry	Scientists develop explanations using observations (evidence) and what they already know about the world (scientific knowledge). Good explanations are based on evidence from investigations.

The unifying concepts and processes are ways of thinking about science in order to understand the natural and designed world in which we live. The *National Science Education Standards* state that "Lifelong scientific literacy begins with attitudes and values established in the earliest years." Therefore, the teacher of young children has the important task of providing the foundation of scientific literacy by guiding students to do critical and creative thinking about the way things work. Questioning strategies designed to engage young students in this type of thinking are provided in **Constancy, Change, and Measurement** at the end of each activity description. Although most things are in the process of change, some properties of objects can be characterized by constancy. Measurement provides a quantitative way to describe both constancy and change in the characteristics of organisms; for example, the length, height, width, volume, mass, and temperature can be measured to describe characteristics of organisms and their environments.

Big Ideas about Seeds:
- Seeds have characteristics/properties.
- Every seed has a little plant inside with food all around it.
- A seed can grow under favorable conditions.

Materials: color crayons for each student, small objects you and your students have collected including seeds such as popcorn kernels and 15-bean soup mix, and objects that are not seeds such as cinnamon candy drops, gravel, and cereal

Teaching Tips: Have students divide the small objects they have collected into two groups: seeds and not seeds. Ask them to trace around each object and color it as accurately as possible. Encourage students to ask their own questions: "What are these? How can we tell which ones are seeds and which ones are not seeds? Will they change? Will they grow?" Students may indicate that color or texture make a seed a seed; however, point out more helpful criteria such as "a little mark" on seeds if someone mentions the scar on a seed. Students themselves will be highly critical of one another's reasons for calling an object a seed. Try to promote such critical interplay among them whenever possible. From time to time, you will need to take part in their discussions yourself. Try to point out inconsistencies in their arguments. If a student says that an object is a seed because it's bumpy, you could point to a bumpy object in her group of NOT SEEDS; if she says that something can't be a seed because it's too shiny, show her a shiny object in her SEED group. Have students circle the color words that describe the seeds they drew. Ask students how they feel about the smell of seeds and color in the face that best expresses their feeling. Then ask students how they could find out which ones are really seeds. Have students write a response or draw a diagram to show their ideas.

Connect to Content

Inquiry/Characteristics of Organisms: Investigations for primary students are largely based on systematic observations to differentiate between living and nonliving. Primary students associate "living" with any object that is active. This view of living develops into one in which movement becomes the defining characteristic. Life science in grades K-4 should provide opportunities for students to increase their understanding of the characteristics of organisms that they encounter in their everyday lives. Young children begin their study of organisms by examining and qualitatively describing organisms and their behavior. Later on, children combine other concepts, such as eating, breathing, and reproducing to define life. When carefully observed and described, the characteristics of organisms and changes in characteristics over time, provide the necessary precursors to the later introduction of more abstract ideas.

Seeds are more than just a way to grow food - many seeds are food! For example, some seeds are used to season food, such as anise, caraway, coriander, dill, pepper, and celery seeds. Others, like corn and peanuts, are squeezed to make oil. Still others make great snacks when roasted; for example, pumpkin and sunflower seeds.

Constancy, Change, and Measurement: Ask: Is there anything (characteristic) that all seeds have in common? (Constancy - scar, baby plant or embryo, food for developing plant, and a seed coat) How do seeds change? (Change - grows into a plant, develops root system and shoot system) How could seeds be measured? (Measurement - length, mass, volume, etc.)

Observe Seeds Scoring Rubric

4 points   correct, complete, detailed
3 points   partially correct, complete, detailed
2 points   partially correct, partially complete, lacks some detail
1 point   incorrect or incomplete, needs assistance

Scoring Criteria	4	3	2	1
Followed directions and completed activity.				
Drew "seeds" and" not seeds".				
Identified colors of seeds.				
Identified how seeds smell by coloring in appropriate face.				
Suggested a way to find out which objects are seeds in words or drawings.				

Total Points_____

Communicate about Seeds

Materials: dried lima beans (approximately 100), color crayons for each student, hand lens

Teaching Tips: Twenty-four hours before doing this activity, put 50 dried lima beans into a container; cover with bleach solution (1 liter of water and 1/2 teaspoon of bleach), and let them soak. If you soak them much longer than this, they will start to mold. Questions such as these will guide the students' observations: "How are the dry seed and wet seed different? Is the size different? Is the outside of the seed different?" Some students may suggest that the outside skin of the soaked seed pops off and there are two pieces inside. Help students observe the tiny embryo or baby plant with its root and leaves at one end of the seed. You may have to assist students to see the baby plant with a hand lens because everything will be white.

Connect to Content

Inquiry/Characteristics of Organisms: A seed makes new plants and contains a baby plant hidden inside. At one end of a seed there is a little bump. That's the baby plant, called an embryo. If you examine the bump with a hand lens, you may be able to see tiny leaves sticking out. The rest of the seed is stored food that the baby plant uses as it starts to grow. In the early stages of growth, water enters the seed, causing it to swell and splitting the outer seed coat.

Next, roots emerge from the seed and grow downward, and finally the stem and leaves (the shoot) appear and grow upward. Water softens the food stored in the seed. The food is broken down and dissolved. The seedling uses this dissolved food to germinate.

Constancy, Change, and Measurement: Ask: How did the wet seed change? (Change - swollen, seed coat split - in time, the root will emerge and the stem and leaves (shoot) will appear)

Communicate about Seeds Scoring Rubric

4 points   correct, complete, detailed
3 points   partially correct, complete, detailed
2 points   partially correct, partially complete, lacks some detail
1 point   incorrect or incomplete, needs assistance

Scoring Criteria	4	3	2	1
Followed directions and completed activity.				
Recorded observations of dry seed using words and a drawing.				
Recorded observations of closed wet seed using words and a drawing.				
Recorded observations of both sides of the open wet seed using drawings.				
Described the inside of the seed.				

Total Points_____

Classify Seeds

Materials: a variety of seeds such as garbonzo beans, peas (from a pod), popcorn, lima beans (some grocery stores have bulk food sections with interesting beans or 15-bean soup mix can be a good source for a variety of beans), white glue or color crayons for each student

Teaching Tips: Try to select seeds for this activity such as garbonzo beans for bumpy seeds; peas from a pod for soft seeds; popcorn for hard seeds or pointy seeds; lima beans for smooth seeds or round ends, etc. Ask students to touch the seeds. Then prompt students to find a bumpy seed and glue it in the box. Continue asking students to find seeds with the following properties: soft, hard, rough, rounded, pointed, big, small, shiny, flat, dark, light, and dull. When students have identified a seed with the property, have them glue to seed in the appropriate box.

Invite students to complete the following sentences using words from the data sheet: Seeds can feel _____.

Ask students questions to discuss the idea that seeds have multiple properties; say, "Can bumpy seeds like garbonzo beans be smooth or rough?" Point out that a seed may have several of the properties listed and may be classified in more than one category.

Connect to Content

Inquiry/Characteristics of Organisms: Primary students generally use mutually exclusive rather than hierarchical categories when classifying. Students do not consistently use classification schemes similar to those used by biologists until the upper elementary grades. A seed plant develops into an embryo that is packaged along with a food supply within a seed coat. This protects the dormant embryo from drought, cold, and other harsh conditions. Seeds may differ in size, shape, color, and texture but all seeds have the potential to develop into new plants.

Constancy, Change, and Measurement: Ask," Is there one way that all of these seeds are alike?" (Constancy - all will develop into new plants)

Classify Seeds Scoring Rubric

4 points   correct, complete, detailed
3 points   partially correct, complete, detailed
2 points   partially correct, partially complete, lacks some detail
1 point   incorrect or incomplete, needs assistance

Scoring Criteria	4	3	2	1
Followed directions and completed activity.				
Used senses to observe seeds.				
Glued or drew a seed for each characteristic.				
Sorted seeds accurately for each characteristic.				

Total Points_____

Measure with Seeds

Materials: For each student: one craft stick (Popsicle stick), white glue, 10 seeds that are 1 centimeter or longer, assorted classroom objects to measure

Teaching Tips: Use seeds that are 1 centimeter or longer (lima beans, acorns, avocado pits, peach pits, nuts, pumpkin seeds). Ask students to pick one seed and trace it on the ruler. Then challenge students to find 9 more seed that are the same size by placing the seeds on the tracing to compare the size. Have students glue their 10 seeds to a craft stick. You will have to interrupt this activity to allow the white glue to dry. When the glue is dry demonstrate how to use the seed stick to measure the length of the assorted classroom objects listed on the activity sheet. Then have students select classroom objects to measure with their seed sticks. Ask student why the measurements are different although students may have measured the same object. The seeds may be different sizes. You may wish to discuss the need for standardized measuring tools such as rulers and the need to list the unit of measurement. A standardized measuring tool or unit provides the same frame of reference for everyone.

Connect to Content

Inquiry/Characteristics of Organisms: The seeds are dormant to increase the chances that germination will occur at a time and place most advantageous to the seedling. Seeds of desert plants, for example, germinate only after a substantial rainfall. If they were to germinate after a modest drizzle, the soil might soon be too dry to support the seedlings. Where natural fires are common, many seeds require intense heat to break dormancy, so that seedlings are most abundant after fire has cleared away competing vegetation. Where winters are harsh, seeds may require extended exposure to cold; seeds sown during summer or fall do not germinate until the following spring. Very small seeds, such as those of some lettuce varieties, require light for germination and will break dormancy only if they are buried shallow enough for the seedlings to poke through the soil surface. Some seeds have coats that must be weakened by chemical attack as they pass through an animal's digestive tract and thus are likely to be carried some distance before germinating.

Constancy, Change, and Measurement: Ask: Why do you think some of our measurements are different for the same object? The seeds may be different sizes. Discuss the need for standardized measuring tools such as rulers and the need to always name the unit of measurement. Point out that a standardized measuring tool or unit provides the same frame of reference for everyone.

Measuring with Seeds Scoring Rubric

4 points   correct, complete, detailed
3 points   partially correct, complete, detailed
2 points   partially correct, partially complete, lacks some detail
1 point   incorrect or incomplete, needs assistance

Scoring Criteria	4	3	2	1
Followed directions and completed activity.				
Constructed a seed stick.				
Measured the objects listed on the activity sheet with the seed stick.				
Measured 5 additional objects with the seed stick.				
Wrote a statement indicating that the seeds were not a standard unit of measurement.				

Total Points_____

Infer about Seeds

Materials: For each group of 4 students: apple, green bean, plum, corn in husk, a knife for the teacher to cut the fruit, paper plates

Teaching Tips: Give each group the set of 4 fruits (apple, green bean, plum, corn in husk) on a paper plate and identify them as fruits. Ask the students what they might find inside each fruit. Have them draw a picture to show the seeds inside each fruit. Then cut the fruit open and give groups of students the opened fruit to observe. Have students draw pictures to show what the seeds actually look like in each fruit. Ask students to describe and compare the seeds, "Are the seed small or large, round or flat, rough or smooth, hard or soft? Which fruit has the most seeds? Which fruit had the fewest seeds? How are the seeds alike? How are the seed different?" You may want to plant the seeds and observe the new plants. In a class discussion, develop a characteristic list for fruits describing both the outside of the fruits and the seeds inside. Then have students compare the seeds and write a statement about what they found out.

Connect to Content

Inquiry/Characteristics of Organisms: Fruit is the name given to the structure that develops from the flower of a plant and contains seeds. Although we consider beans and corn to be vegetables, botanically speaking, they are all fruits because they contain seeds. Fruits provide us with tasty food to eat and provide the embryo (young plant) with protection and a mechanism for dispersal. Fruits are classified into several types, depending on their origin such as simple fruits which develop from a single ovary. Simple fruits may be fleshy, such as an apple, or dry, such as a soybean pod. A strawberry is an aggregate fruit resulting from a single flower that has several separate carpels. A pineapple is a multiple fruit developing from a group of separate flowers tightly clustered together.

Constancy, Change, and Measurement: Ask: How are all of these fruits alike? (Constancy - fruits all contain seeds)

Inferring about Seeds Scoring Rubric

4 points   correct, complete, detailed
3 points   partially correct, complete, detailed
2 points   partially correct, partially complete, lacks some detail
1 point   incorrect or incomplete, needs assistance

Scoring Criteria	4	3	2	1
Followed directions and completed activity.				
Recorded inferences about the seeds inside each fruit with a drawing.				
Recorded how the seeds actually looked with a drawing.				
Wrote a statement about what was found out when comparing the seeds.				

Total Points _____

Predict with Seeds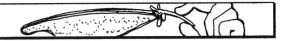

Materials: seed pods (milkweed pods, pea pods, bean pods, peanuts, etc.) - 1 per student

Teaching Tips: Have students look at the seed pod. Prompt them to feel the seed pod. Then ask students to trace and color the seed pod. Ask students to predict how many seeds are inside. Then have them open the pod and count the seeds. Ask them to trace and color the open seed pod. Have students report how many seeds were in the pod. Make a class graph to show how many seeds each person found in a pod. This type of graph is a histogram. Use Post-it® notes to indicate each person that had 2 seeds in a pod, 3 seeds in a pod, etc.

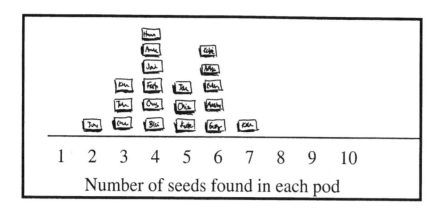

Number of seeds found in each pod

Then ask:
What was the most number of seeds found in a pod?
What was the least number of seeds found in a pod?
What number of seeds did most people find in a pod?

Point out the little stem with which each seed is fastened to the pod. Students may associate it with the scars they observed on some of the seeds they looked at before. Introduce the idea that many kinds of seeds are enclosed by a pod or some other type of cover. Ask, "Can you think of any seed coverings? What seed coverings do your eat? How do you think beans and peas escape from their covers? If nobody every picks the pods off the plant, what happens to the seeds? Where have you found empty seed pods? Where did the seeds go? Have you every found seeds clinging to your clothes or to your pets?"

Connect to Content

Inquiry/Characteristics of Organisms: Histograms record the number of times something occurs or frequency of events. Introduce the histogram as a method of recording information. Legumes bear their seeds in pods (shells). And like all seeds, they can grow into a new plant. Edible legumes include lima beans, peas, snap beans, peanuts, and soybeans. The legumes or pulse crops can form a symbiotic relationship with *Rhizobium sp.*, which are nitrogen-fixing bacteria. Of all mineral elements, nitrogen is the one that most often limits the growth of plants and yields of crops. Plants require nitrogen and although the atmosphere is nearly 80% nitrogen, plants can't use nitrogen in that form. Nitrogen-fixing bacteria restock nitrogenous minerals which plants can absorb in the soil by a metabolic process called nitrogen fixation.

Constancy, Change, and Measurement: Ask: If you opened another pod, how many seeds do you think you might find inside? (Constancy - the number of seeds that occurred the most often on the class histogram would be an accurate prediction)

Predict with Seeds Scoring Rubric

4 points   correct, complete, detailed
3 points   partially correct, complete, detailed
2 points   partially correct, partially complete, lacks some detail
1 point   incorrect or incomplete, needs assistance

Scoring Criteria	4	3	2	1
Followed directions and completed activity.				
Observed, traced, and colored the seed pod.				
Predicted how many seeds were in the pod.				
Observed, traced, and colored the opened seed pod.				
Recorded the most, least, and most frequent number of seed found based on class data chart.				

Total Points_____

Making Model Seeds

Materials: a variety of seeds that are dispersed by air or water, by sticking to animals, and by being eaten by animals or pictures of seeds from books in the library, a variety of materials that students can use to construct model seeds   (Optional: 1 pair of old wool socks for each student and a pair of tweezers)

Teaching Tips: If you prefer to have students collect seeds, have them put an old pair of wool socks over their shoes. Then take them on a walk through a vacant lot or woods. Explain that hitchhiking seeds will hang on to you just like they do on an animal's coat. Take the socks off and pick the seeds off with tweezers. Ask students how seeds that travel by air are alike and how they are different from seeds that travel by other mechanisms. Continue with similar questions about seeds that hitchhike, are eaten, and travel by water. Have students describe and demonstrate how the shape of the seeds helps them travel. Then have them draw a picture of a seed and tell how it travels. Collect a variety of materials and make them available for students to use to make a physical model of their seed.

Connect to Content

Inquiry/Characteristics of Organisms: Seeds have to be ingenious to get around because they don't have legs! Animals and birds often unknowingly help seeds spread. For example, when squirrels store nuts for future meals, they sometimes forget where they bury them, leaving a trail of trees and plants to mark their absent-mindedness. When birds eat berries, they can't digest the seeds and excrete them after flying far away from the original plant. Some seeds are "hitchhikers." They have small hooks or barbs to attach themselves to any animal - including you - that comes by. They get carried off to other locations and dropped there. "Parachuters," such as dandelions, and "winged" seeds, such as those from maple trees, are so light that they are blown easily from place to place. "Shooting" seeds are formed in pods that burst open and fire them out.

Constancy, Change, and Measurement: Ask: What do all "hitchhiker" seeds need? (Constancy - need something that allows them to attach to animals) What do all seeds that travel by air need? (Constancy - need to be light and have a structure that allows them to catch the wind)

Making Model Seeds Scoring Rubric

4 points   correct, complete, detailed
3 points   partially correct, complete, detailed
2 points   partially correct, partially complete, lacks some detail
1 point   incorrect or incomplete, needs assistance

Scoring Criteria	4	3	2	1
Followed directions and completed activity.				
Drew a picture of a model seed.				
Described how the model seed travels.				
Method of travel and characteristics of seed matched.				
Constructed a physical model of a seed.				

Total Points_____

Defining Operationally: What are Seeds?

Materials: soaked lima beans - put in a container, cover with weak bleach solution (1 liter of water and 1/2 teaspoon of bleach), soak for 24 hours

Teaching Tips: You may have students reflect on their experiences with seeds or have them germinate seeds in a Zip-lock® bag using the following steps:
1. Wet a paper towel and put it in a Zip-lock® bag.
2. Put 4 seeds in the bag.
   • Seal the bag!
   • Put you name on the bag.
3. Observe the seeds each day.
4. Draw a picture of one of the seeds each day to show changes.
Have students put 4 seeds in to bag in case some do not germinate. Suggest that students draw a picture of the seed that is changing the most each day.

Use a Sharpie® pen to put each student's name on his/her bag. Have students examine the germinating seeds from day to day. They will gradually build a definition of what a seed is in terms of what a seed does. Students may notice a "little wormy thing." Ask, "What do you think it might be? Is it growing? What do you think it will grow into?"

Have students draw or write responses to each item:
Seeds look _____. (shape, color, size, etc.)
Seeds have _____. (baby plant - embryo, food supply, seed coat)
Seeds can grow if _____. (proper temperature, sufficient moisture, and oxygen)
Seeds come from _____. (mature plants)
Then have students put all of these ideas together to complete their operational definition of a seed.
A seed is ___(A seed is a plant waiting to start growth and development.)___

Connect to Content

Inquiry/Characteristics of Organisms: Seeds change very fast in the zip-lock® bag. First the seeds swell and the skin breaks. In a few days, roots push out from one end. Soon after that, stems sprout from the other end. Now the baby plants are seedlings. The seedlings get along without soil because they are living on the food stored in the seeds. As water soaks into the seed, the food is dissolved, broken down, and flows into the new roots and stems, bringing them everything they need to grow. Most germinating seeds do not need light to sprout. The bean plant always sprouts a single thick root first.

Constancy, Change, and Measurement: Ask: How did your seeds change? (Change - seed became swollen, seed spit, roots emerged, stem and leaves appeared) How could you measure that change? (Measurement - use ruler to measure length of seed, root, shoot - stem and leaves)

What is a Seed? Scoring Rubric

4 points   correct, complete, detailed
3 points   partially correct, complete, detailed
2 points   partially correct, partially complete, lacks some detail
1 point   incorrect or incomplete, needs assistance

Scoring Criteria	4	3	2	1
Followed directions and completed activity.				
Drew or wrote a description of how a seed looks.				
Drew or wrote what seeds have.				
Drew or wrote what seeds need to grow.				
Drew or wrote where seeds come from.				
Synthesized information to write a definition of a seed.				

Total Points_____

Investigate Seeds

Materials: 50 waxed paper cups or cut down milk cartons, collection of seeds, potting soil, 10 sheets of construction paper cut into strips about 1 centimeters wide

Teaching Tips: Have each student select a different kind of seed to plant. Instruct students to tape one of their seeds to the outside of a cup in order to identify the type of seed planted in the cup. Then have them fill the cup with potting soil. Demonstrate how to poke 3 holes in the soil with a finger. Then show students how to place a seed in each hole. Explain to students that they should put a little soil over each seed. Then show students how to pour water in the cup. Cut 10 sheets of construction paper into strips about 1 centimeters wide. Have students measure the plants each day by cutting strips of paper the same height as the plants. Have them glue the strips to the back of their paper or on a large piece of paper each day. Ask the following questions:

What day did your plant grow the most?

When did it grow the least?

Do all seeds grow the same?

The most frequent source of trouble in growing seeds is molding. This is usually caused by over-watering. Puncture the bottoms of the containers to allow for some drainage. Weekends and short vacations may pose the opposite problem - since the containers are small, they may dry out in a few days. A plastic bag or plastic wrap placed around the plant makes a little greenhouse to prevent rapid loss of moisture, and also to protect from excessive cold.

Connect to Content

Inquiry/Characteristics of Organisms: The seeds are dormant to increase the chances that germination will occur at a time and place most advantageous to the seedling. Seeds of desert plants, for example, germinate only after a substantial rainfall. If they were to germinate after a modest drizzle, the soil might soon be too dry to support the seedlings. Where natural fires are common, many seeds require intense heat to break dormancy, so that seedlings are most abundant after fire has cleared away competing vegetation. Where winters are harsh, seeds may require extended exposure to cold; seeds sown during summer or fall do not germinate until the following spring. Very small seeds, such as those of some lettuce varieties, require light for germination and will break dormancy only if they are buried shallow enough for the seedlings to poke through the soil surface. Some seeds have coats that must be weakened by chemical attack as they pass through an animal's digestive tract and thus are likely to be carried some distance before germinating.

Constancy, Change, and Measurement: Ask: What day did your plant grow the most? (Change, measurement) When did it grow the least? (Change, measurement) Do all seeds grow the same? (Constancy)

Observe Seeds Scoring Rubric

4 points   correct, complete, detailed
3 points   partially correct, complete, detailed
2 points   partially correct, partially complete, lacks some detail
1 point   incorrect or incomplete, needs assistance

Scoring Criteria	4	3	2	1
Followed directions and completed activity.				
Measured plant's growth each day.				
Cut strips of paper the same height as the plant each day.				
Glued strips on paper to make a graph.				
Wrote what was found out in the investigation				

Total Points_____

Observe Seeds

1. Observe your group of objects.
2. Which are seeds?
3. Draw the objects in the boxes.

Seeds	Not Seeds

4. What colors are the seeds? yellow, orange, red, purple, blue, green, white, black

5. How do the seeds smell?

6. How can you find out which ones really are seeds?

Communicate about **Seeds**

1. Observe a dry seed.

Tell: _____

_____

_____

Draw

2. Observe a wet seed.

Tell: _____

_____

_____

Draw

3. Open the wet seed. Observe both sides.

Draw

Draw

4. Tell about the inside of the seed. _____

_____

_____

Classify Seeds

Observe your pile of seeds. Use your senses to sort your seeds into groups. Find at least one seed for each box. Glue or color.

Touch

Smooth

Bumpy

Soft

Hard

Rough

Rounded

Pointed

See

Big

Small

Shiny

Flat

Dark

Light

Dull

Measure with

1. Pick one seed.
Trace it on the ruler.

| | | | | | | |
|1|2|3|4|5|6|7|

2. Pick 9 more seeds the same size.

3. Make a seed stick.
Glue 10 seeds to a craft stick.

4. Use your seed stick to measure.

Name	Number of Sticks	Name	Number of Sticks
book			
desk			
foot			
hand			
height			

5. Why do you think measurements are different?

Infer about Seeds

1. Observe the 4 fruits.

2. Draw a picture to show what the seeds look like inside each fruit.

Apple	Greenbean	Plum	Corn

3. Draw a picture to show what the seeds look like inside the open fruits.

Apple	Greenbean	Plum	Corn

4. Compare the seeds. Tell what you found out.

_____

_____

Predict with Seeds

1. Observe the seed pod.
Trace and color the seed pod.

2. Predict how many seeds are inside.
? I think there are ☐ seeds.

3. Open the seed pod. I count ☐ seeds.
Trace and color the open seed pod.

4. Compare your count.
The most seeds found: ☐
The least seeds found: ☐
The number most found: ☐

Making Model Seeds

• Seeds travel away from their parent plants.

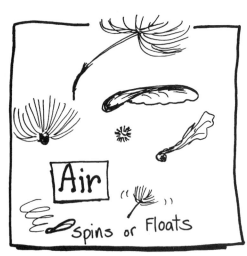

Air
Spins or Floats

Hitchhike
sticks to you

Eaten

Water
floats

1. Make a model of a seed that can travel.

2. Draw a picture of your model.

3. Tell how your model seed travels. _____

_____

What are **Seeds** ?

Defining Operationally

You have studied seeds. Draw or write what you know.

1. Seeds look 👀

2. Seeds have

3. Seeds can grow if

4. Seeds come from

5. Put all your ideas together. Write a rule for a seed.

A seed is_____

_____

_____

_____

Investigate Seeds

1. Plant a seed in a cup of soil.

2. Add water and light.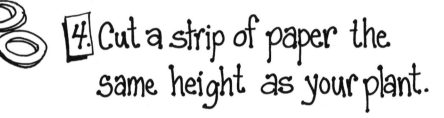
tape a seed →

3. Measure your plant's growth each day.

4. Cut a strip of paper the same height as your plant.

5. Glue the strips of paper to the back of this page.

6. Tell what you found out.

MATTER

These activities were designed with candy in mind because candy is a highly motivating manipulative for young children (and the authors, too). However, candy does NOT have to be used in all of the activities; if you do not feel it is appropriate to use candy or sugar free candy (Jolly Ranchers®, etc.) with your students, you may substitute assorted nuts, beans, fruits, dried fruits, trail mix, vegetables, cereal, etc. Use these activity sheets to engage your students in using the processes of inquiry to discover properties of objects and materials with any items that you determine are easily available, accessible, and appropriate. We have attempted to capitalize on students' natural interest in candy in order to go beyond just the enjoyment of eating it!

National Science Education Standards

Unifying Concepts and Processes: Constancy, Change, and Measurement	
Properties of Objects and Materials	Objects have many observable properties, including size, weight, shape, color, temperature, and the ability to react with other substances.
Abilities Necessary to do Scientific Inquiry	As students focus on the processes of inquiry, they develop the ability to ask scientific questions, investigate aspects of the world around them, and use their observations to construct reasonable explanations for the questions posed.
Understanding About Scientific Inquiry	Scientists develop explanations using observations (evidence) and what they already know about the world (scientific knowledge). Good explanations are based on evidence from investigations.

The unifying concepts and processes are ways of thinking about science in order to understand the natural and designed world in which we live. The *National Science Education Standards* state that "Lifelong scientific literacy begins with attitudes and values established in the earliest years." Therefore, the teacher of young children has the important task of providing the foundation of scientific literacy by guiding students to do critical and creative thinking about the way things work. Questioning strategies designed to engage young students in this type of thinking are provided in **Constancy, Change, and Measurement** at the end of each activity description. Although most things are in the process of change, some properties of objects can be characterized by constancy. Measurement provides a quantitative way to describe both constancy and change in the properties of objects and materials; for example, the length, height, width, volume, mass, and temperature can be measured to describe objects and materials.

Materials: 1 Gobstopper® for each student, clear plastic cup half-filled with water [Note: If you are not able to find Gobstoppers®, you can use other hard candies that have a brightly colored coating.]

Teaching Tips: Enter the "Wild World of Wonka" at www.wonka.com to check out the Willy Wonka Candy Factory. Explain to students that they will be observing changes in a solid (hold up a Gobstopper®) and a liquid, water. Tell students that once the Gobstopper® is placed in water, it is very important that the cup containing the Gobstopper® and the water be kept perfectly still. Instruct students to carefully observe the container and draw the changes they observe on their data sheets. Guide observations by asking questions such as, "What is happening to the Gobstopper®? What is happening to the water?" You may want to watch a Gobstopper® and when it changes color, ask <u>all</u> students to draw what they see in their container on their data sheets. When the Gobstopper® has changed color again, direct <u>all</u> students to go to the next diagram and draw what they see in the container, etc. If the containers, are not moved, concentric circles of color will appear as the pigments on the Gobstopper® are dispersed by the water. The colors will be on the bottom of the cup because the pigments are denser than the water. Students may be able to see a clear area between each concentric circle, this is dextrose sugar which is used to seal in each color/flavor.
Patterns of change that can be expected for each Gobstopper®:
Purple -> red -> white -> pink/purple
Red -> yellow -> white -> pink/purple
Yellow -> orange -> green -> pink/purple
Green -> yellow -> orange -> pink/purple
Orange -> yellow -> green -> white -> pink/purple

Connect to Content

Inquiry/Properties of Objects and Materials: Investigations for primary students are largely based on systematic observations. Physical science in grades K-4 should provide opportunities for students to increase their understanding of the characteristics of objects and materials that they encounter in their everyday lives. Young children begin their study of matter by examining and qualitatively describing objects and their behavior. When carefully observed and described, the properties of objects and materials, changes in properties over time, and the changes that occur when materials interact provide the necessary precursors to the later introduction of more abstract ideas.

Gobstoppers®: The dye that gives Gobstoppers® their distinct appearance begins as a solution because moisture is needed to develop the color in the dyes. The dye solution is placed in a spray dryer which uses air to drive off the moisture resulting in dry tablets called *lakes*. These lakes are then attached to aluminum and are termed *aluminum lakes*. The aluminum lakes are not water soluble but are water dispersible. It is these dispersions of color that can be seen when the Gobstoppers® are placed in water. The aluminum lakes also cause the spectacular patterns observed in the container as the dyes sink in the water due to the fact that they are denser than water. Each dye and flavoring is sealed with several coats of dextrose syrup. Since dextrose, a sugar, dissolves in water, there is a clear area between each color in the water where the sugar has dissolved. The candy is polished with a thin coat of carnauba wax (the same wax used for polishing your car or furniture that comes from palm tree leaves and buds) which can be seen on the surface of the water.

Constancy, Change, and Measurement: Ask: How did the red (orange, yellow, green, purple) Gobstopper® change? Was the pattern of change the same (constant) for the red (orange, yellow, green, purple) Gobstopper®? How did the water change? Was the pattern of change the same (constant)? What are some ways that we can measure the changes you observed?

Although the Gobstoppers® are changing, some properties are characterized by constancy, including the sequence of colors on each kind of Gobstopper®. Changes in the Gobstoppers® can be measured; for example the time it takes for each change in the Gobstopper® and/or water.

Observe Matter Scoring Rubric

4 points   correct, complete, detailed
3 points   partially correct, complete, detailed
2 points   partially correct, partially complete, lacks some detail
1 point   incorrect or incomplete, needs assistance

Scoring Criteria	4	3	2	1
Followed directions and completed activity.				
Recorded initial color of candy and water.				
Recorded second color of candy and changes in water.				
Recorded third color of candy and changes in water.				
Recorded fourth color of candy and changes in water.				

Total Points_____

Materials: crayons and one square Post-It® notes for each student, box or bag of assorted candy such as M & Ms®, Skittles®, Gobstoppers®, Life Savers®, Runts®, etc. for the class [Note: If you do not want to use candy or sugar free candy (Jolly Ranchers®, etc.), you may substitute assorted nuts, beans, fruits, dried fruits, trail mix, vegetables, cereal, etc.]

Teaching Tips: Enter the "Wild World of Wonka" at www.wonka.com to check out the Willy Wonka Candy Factory or find out about Life Savers® at www.CANDYSTAND.com. You may use one bag or box of candy (or other item) for the class. Give each student one of the pieces. Ask students to draw a picture of it on a Post-It® note. Then have one student describe his/her piece. Ask if any other students have a piece that matches the description. Have students hold up their piece if it matches. Then ask if anyone has a piece that is different from the first piece. Have the student tell how the piece is different and ask other students that have the same kind to hold their pieces up. Continue this until all the different kinds have been described. Then ask questions such as, "Which kind did the greatest number of children in class have? Which kind did the least number of children in class have?" Elicit suggestions from the students about how they could tell for sure. If it isn't suggested, propose that they class could construct a graph to show how many people had each kind of piece. Begin by labeling the different kinds of pieces the students have below a horizontal line drawn on a chalkboard. One by one have each student bring the Post-It® note with a drawing of his/her piece and construct a graph on the chalkboard by placing the Post-It® note in the column labeled for each kind of piece. You may have to guide students to place the first Post-It® on the horizontal line and then place the next Post-It on top of the first square. Each time a student places a Post-It® note on the chalkboard, have students color in their data sheets to correspond with the class graph. Students can color the graph boxes the same color as the piece.

Connect to Content

Inquiry/Properties of Objects and Materials: Children's natural curiosity leads to explorations of the world around them as they observe and manipulate common objects and materials. As children observe and describe these objects and materials, they begin to form explanations of the world. Experiences that provide opportunities for children to observe and describe help children develop a subject-matter knowledge base in order to give explanations and make predictions about objects and events in the world. Children reflect on the similarities and differences of objects as they observe, manipulate, and classify common objects. As a result, their initial sketches and single-word descriptions lead to increasingly more detailed drawings and richer verbal descriptions.

Constancy, Change, and Measurement: Ask: What is a true statement that you can make by looking at the graph? Did everyone have the same kind of candy? (Constancy) How many different kinds of candy did the children in our class have? (Change)

How many more children had _____ than _____? (Measurement)
How many fewer children had _____ than _____? (Measurement)

Communicate Matter Scoring Rubric

4 points   correct, complete, detailed
3 points   partially correct, complete, detailed
2 points   partially correct, partially complete, lacks some detail
1 point   incorrect or incomplete, needs assistance

Scoring Criteria	4	3	2	1
Followed directions and completed activity.				
Colored the graph to show the number of candies.				
Identified which color had the greatest number of candies.				
Identified which color had the fewest number of candies.				
Identified which colors had the same number of candies.				
Wrote a sentence that analyzed and/or synthesized the information on graph.				

Total Points _____

Classify Matter

Materials: crayons for each student, a gallon-sized Zip-lock® bag of assorted candy such as M & Ms®, Skittles®, Gobstoppers®, Life Savers®, Runts®, Jelly Bellys® etc. for each group. You can also use individually wrapped candy and eliminate the need for a gallon-sized Zip-lock® bag.
[Note: If you do not want to use candy or sugar free candy (Jolly Ranchers®, etc.), you may substitute assorted nuts, beans, fruits, dried fruits, trail mix, vegetables, cereal, etc.]
6 circles (Zip-lock® for candies if not individually wrapped)

Teaching Tips: Enter the "Wild World of Wonka" at www.wonka.com to check out the Willy Wonka Candy Factory or find out about Life Savers® at www.CANDYSTAND.com. Give each group a gallon-sized Zip-lock® bag of 20-30 assorted candies or other items or individually wrapped candies. Tell students that they are not to open the Zip-lock® bag and demonstrate how they can move the items in the Zip-lock® bag by pushing the candies around through the bag (this keeps unwrapped candy free from contaminants that might be on hands). Ask students to observe and describe the candies or other items. Ask students how the candies are alike and different. Then ask students to sort the candies into groups based on a property of the candy such as color, shape, size, etc. Show students how to place the candies into the circles on the data sheet and trace around each one. Tell students to label each group to show why the items were sorted into the same set. Ask students if they can think of another way to sort their candy. Have them list one other way to sort their objects.

Connect to Content:

Inquiry/Properties of Objects and Materials: As children compare, describe, and sort objects, they begin to form explanations about the world and the way it works. Through the observation, manipulation, and classification of common objects, children reflect on the similarities and differences of the objects. As a result, their initial sketches and single-word descriptions lead to increasingly more detailed drawings and richer verbal descriptions.

Constancy, Change, and Measurement: Ask: Can you think of one way that all candy is alike? (Constancy - students may suggest that all candy tastes "sweet" or "good") What are some of the ways that candy differs? (Change - taste, texture, size, shape)

Classify Matter Scoring Rubric

4 points   correct, complete, detailed
3 points   partially correct, complete, detailed
2 points   partially correct, partially complete, lacks some detail
1 point   incorrect or incomplete, needs assistance

Scoring Criteria	4	3	2	1
Followed directions and completed activity.				
Sorted objects into groups based on one property.				
Drew and colored objects to pictorially represent classification scheme.				
Labeled each group with the property used to sort the objects.				

Total Points   _____

Measure Matter

Materials: one shoe box lid, one film canister, one toilet paper tube (mystery tube), three plastic cups each filled with a different sized individually wrapped candy such Life Savers®, or Jolly Ranchers® (small, medium, and large)
[Note: If you do not want to use candy or sugar free candy (Jolly Ranchers®, etc.), you may substitute assorted nuts, beans, fruits, dried fruits, trail mix, vegetables, cereal, etc.]

Teaching Tips: Have students observe and describe the objects in the cups. Ask them to put the containers in order according to size of the objects in each cup. Ask students to estimate how many pieces of the medium-sized objects will fit into the film canister. Have them record their estimate on their data sheets. Instruct students to put their film canister on the box lid. Then have students place the mystery tube (toilet paper tube) over the film canister so that they can't see the film canister. Tell students to count out the objects they estimated would fill the film canister. Have them drop their objects into the mystery tube. Then have them carefully pull the mystery tube up and observe if some of the objects fall into the lid or if the objects did not fill the tube. Ask, "How many spilled out?" and" How many more do you have to add to fill the film canister?" Have students record this information on their data sheets. Then ask, "How many pieces of medium-sized candy will film canister hold?" Repeat having students estimate, count out pieces, drop them into the mystery tube, pull up the mystery tube, find out if they estimated too many or too few, and record their data.

Connect to Content:

Inquiry/Properties of Objects and Materials: As children develop expertise with language, their descriptions become richer and include more detail. At first, descriptions of objects will be qualitative (words) but eventually children learn that they can add to their descriptions by using quantitative (numbers) information acquired by measuring objects. Before measuring tools are introduced, students can create nonstandard devices for measuring, then the need for standardized measurements can be demonstrated and they can be introduced to conventional measuring instruments such as rulers, balances, and thermometers.

Constancy, Change, and Measurement:
Describe different ways that you could measure the candies?
(Measurement - students may suggest obtaining different measurements
such as mass, volume, length, etc. or using different nonstandard and
standard tools such as a balance, measuring cups, tape measure, etc.)

Measure Matter Scoring Rubric

4 points   correct, complete, detailed
3 points   partially correct, complete, detailed
2 points   partially correct, partially complete, lacks some detail
1 point   incorrect or incomplete, needs assistance

Scoring Criteria	4	3	2	1
Followed directions and completed activity.				
Estimated the number of objects that would fit in the container.				
Counted the number of objects that fell out or were needed to make the container full.				
Measured the volume of the container using three different sized objects to fill the container.				

Total Points_____

Predict Matter

Materials: crayons for each student, a "Peek Box" of Gobstoppers® for
each group
Construct "Peek Boxes." Carefully open the box and remove the
Gobstoppers®. Then cut the corner of the box so that the hole is too small
for a Gobstopper® to roll out. Count out 10 Gobstoppers® to put back
into the box such as 3 red, 2 purple, 2 yellow, 2 orange, and 1 green. Number
each box and make a key so you know the colors of the Gobstoppers® in
each box. You may want to tape the box closed where you opened it [Note:
You may also use small boxes such as those for jewelry and use objects
such as marbles or other round objects.]

Teaching Tips: Tell students that they are going to use their "Peek Boxes"
to try to predict the colors and number of Gobstoppers® (or other
objects) of that color that might be in their box. Explain that each box has
10 Gobstoppers® but the number of Gobstoppers® of each color might be
different in each box. Demonstrate how to roll the Gobstoppers® around
and tilt the box so one is showing in the peek hole of the box. Show students

how to record the color on their data sheet. Tell students that they will take turns in their group rolling the Gobstoppers® around in the box, tilting the box, and recording the color they see on their data sheets until they have completed 10 "Peeks." Then they will try to predict what color each of the 10 Gobstoppers® in the box are by looking at the data they collected in their "Peeks." Set up guidelines for the students such as taking turns and no violent shaking of the box. It is important to point out that predictions are not "incorrect" but that some predictions are more accurate than other predictions. Encourage students to think about how they can make more accurate predictions such as taking more samples by rolling the Gobstoppers® around and completing more "Peeks" or combining the class samples for all the boxes that are the same.

Connect to Content

Inquiry/Properties of Objects and Materials: Student answers will probably differ. If the predicted numbers were close to what they observed in their samples, they will have *confidence* in their prediction. However, if their predictions were not close to what they observed in their samples, they will probably be tempted to try it again. The reason for the difference is that change played a part in which Gobstopper® was selected in the sample. Predictions are forecasts of future observations. Observing, inferring, and predicting are interconnected process skills used to make sense out of the world. Making predictions involves observing and inferring in order to look for patterns that will lead to an accurate forecast of a future observation or event. The following chart points out the distinctions among observations, inferences, and predictions:

Observations	Information directly gathered through one or more of the senses
Inferences	Explanations based on direct observations
Predictions	Expectation of a future observation or event

Children's ideas about the way things work should be constantly reviewed and revised. When children test their predictions, they make more observations that either support or contradict their predictions. When a prediction is supported with observations, then students will have greater confidence in their predictions. However, when observations do not support predictions, then new observations and inferences lead to new predictions. This helps children understand that science is tentative and changing as new observations result from testing predictions.

Constancy, Change, and Measurement: Ask: Why do you think different people observed different Gobstoppers? (Chance played a part in which Gobstopper® you could see.) Why do you think the predictions were different? (The predictions were based on observations of the Gobstoppers.) What do you think you could have done in order to make a better prediction? (I could take more "Peeks," combine data from several people that observed the same box.)

Predict Matter Scoring Rubric

4 points   correct, complete, detailed
3 points   partially correct, complete, detailed
2 points   partially correct, partially complete, lacks some detail
1 point   incorrect or incomplete, needs assistance

Scoring Criteria	4	3	2	1
Followed directions and completed activity.				
Recorded color of candy appearing after each shake of the box (10 times).				
Predicted and recorded the number of candies in the box.				
Counted the candies and recorded the actual number in the box.				

Total Points_____

Infer with Matter

Materials: one piece candy such as a Jelly Belly®, a Life Saver®, a Sugar Free Jolly Rancher®, etc.[Note: If you do not want to use candy or sugar free candy (Jolly Ranchers®, etc.), you may substitute assorted nuts, beans, fruits, dried fruits, trail mix, vegetables, cereal, etc.]

Teaching Tips: Visit the Jelly Belly web site at http://www.jellybelly.com/menus to view the 40 official flavors in the languages of different countries (Brazil, China, France, Germany, Greece, Israel, Japan, Korea, Netherlands, Portugal, Saudi Arabia, Spain, UK, and US). Give each student a Jelly Belly® or other item. Ask students to describe their Jelly Belly®. Instruct students to draw and use words to describe their Jelly Belly on their data sheets. Then ask them to infer what flavor they think their Jelly Belly might be based on their observations. Have them taste their Jelly Belly and record actual flavor.

Connect to Content

Inquiry/Properties of Objects and Materials: Adults have 9,000 taste buds on the surfaces of the tongue, roof of the mouth, and throat. Young children have several taste buds in their cheeks that disappear at adolescence. Each taste bud has tiny receptor cells that transmit four basic flavors through nerves leading to the brain. The taste buds on the tip of the tongue are sensitive to sweetness. Those at the upper front portion of the tongue respond to salty flavors. Along the sides of the tongue are taste buds that react to sour tastes. The taste buds for bitter things are located in the back of the tongue and are 10,000 times more sensitive than sweet taste buds to alert us to poisonous substances. The spectrum of flavors we experience are created by a combination of food temperature, texture, and odor, along with taste-bud sensation. We associate different colors of food with specific tastes based on our prior experience. Inferences are based on indirect observations, i.e., the shape, color, texture, and smell of the food substance; however, until direct observations are made by tasting the food, the flavor can only be inferred.

Constancy, Change, and Measurement: Ask: Did all the red (or other color or shape) ones taste the same? (Constancy) How were the flavors of the red ones (or other color or shape) different? (Change)

Infer with Matter Scoring Rubric

4 points   correct, complete, detailed
3 points   partially correct, complete, detailed
2 points   partially correct, partially complete, lacks some detail
1 point   incorrect or incomplete, needs assistance

Scoring Criteria	4	3	2	1
Followed directions and completed activity.				
Described the candy in drawing and words				
Inferred the flavor of the candy				
Observed and described the actual flavor of the candy				
Explained how the inference and observation of the flavor was surprising				

Total Points _____

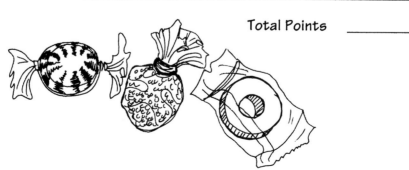

Make a Model of Matter

Materials: a variety of paper, crayons, different types of modeling clay (Model Magic®, etc.) or home-made clay (see recipes below)

Baker's Clay Dough
Ingredients: 1 c. flour, 1/4 c. salt, 1/3 c. water
Mix ingredients together. Knead until smooth. This will need to be baked at 250° - 300°F for one hour *after students make candy model.*

Playdough
Ingredients: 3 c. flour, 1 1/2c. salt, 6 tbsp. oil, 3 tsp. cream of tartar, 3 c. water, food coloring
Mix all ingredients in pan. Heat slowly until mixture "lumps" together. Remove from heat. Stir with a spoon or knead with hands until mixture is the consistency of playdough. Store tightly covered.

Teaching Tips: Have a variety of candies for students to observe and describe. Hold up a candy but put your finger(s) over the name of it (if it is on the wrapper). Have students observe the candy in its wrapper and guess what kind of candy it is. Then ask students why candies have colorful wrappers. Suggest that perhaps the wrappers help to sell the candy. Ask what they look for when purchasing candy. Then tell students that they will have an opportunity to design a candy and wrapper by making a model of it. Gather the ingredients for students to use to make their model of candy. You may want to list the ingredients on chart paper or the chalkboard. Also provide different types of paper that students can use to design and create the candy wrapper. Ask students how their model of candy and wrapper are like real candy and how it differs from the real thing.

Connect to Content

Inquiry/Properties of Objects and Materials:
Models allow children to communicate ideas and explore spatial relationships in three dimensions. Models also provide an opportunity for children to use oral communication as well as written communication such as drawings and language. For young children, a model should be as similar as possible to the actual object or event. A scientific model is a theory formulated to explain and integrate all of the information that is known about a particular natural phenomenon. As new information is discovered, it is tested against the accepted model and if it conflicts, the model or theory may be altered or replaced to accommodate the new data.

Constancy, Change, and Measurement: Ask: How are all of our candy models alike? (Constancy - the same materials used to construct model candy) How are our candy models different from each other? (Change - size, shape, color, etc.)

Make a Model of Matter Scoring Rubric

4 points   correct, complete, detailed
3 points   partially correct, complete, detailed
2 points   partially correct, partially complete, lacks some detail
1 point   incorrect or incomplete, needs assistance

Scoring Criteria	4	3	2	1
Followed directions and completed activity.				
Described ingredients with drawings and/or words				
Made a candy model.				
Made a wrapper for the candy.				
Made a drawing of the candy model				
Made a drawing of the candy wrapper.				

Total Points   _____

Define Operationally - What is Matter?

Materials: ingredients to make "candy" such as the recipe suggested below:

Cereal Balls (10 to 12 balls)
     1/4 cup butter or margarine
     5 cups miniature or 50 large marshmallows
     5 cups favorite ready-to-eat cereal (non-sugared type)
1. Combine butter and marshmallows in large glass mixing bowl.
2. Microwave, uncovered, 2 to 2 1/2 minutes or until melted, stirring once. Stir in cereal, mixing well. Let stand 10 minutes.
3. Form into balls, buttering hands as necessary to prevent sticking. Let stand 1 to 2 hours or until set.

Teaching Tips: Use an easy recipe to make "candy" such as the recipe suggested for *Cereal Balls*. You may want to demonstrate the candy making process and stop at four steps to have students draw a picture of the step you are demonstrating. You may choose to have students actually do the candy making. The decision will depend on your classroom situation. Have students draw four of the steps in proper sequence to illustrate the process of making "candy." Have students taste the candy and indicate how they feel about it by coloring the face that best describes their feelings. Then have students use words or drawings to describe the smell, look, taste, and feel of the "candy."

Connect to Content

Inquiry/Properties of Objects and Materials: When teaching science, children are asked to use their own observations, thoughts, and experiences. However, you, the teacher, have vastly more experience and background to draw upon which presents a major pitfall in deciding when the children's definitions are adequate. A definition that seems adequate to a child, at least according to his/her experiences and observations, may NOT be adequate to you. To correct the child by requiring her/him to use your definition, which is more adequate, from your point of view, but not based on the children's experience would be to defeat a major aspect of the process of inquiry. As a child continues to learn science s/he will meet progressively more complex terms which need to be defined. S/he will also need to revise terms previously defined so as to improve definitions according to his/her recent experiences. To the child, an operational definition can only be stated in terms of activities s/he has done using words, ideas, skills, and operations which s/he has mastered - a *doing* definition. This skill is one to be developed gradually through the child's continuing experience as s/he acquires a higher level of competence in this skill.

Constancy, Change, and Measurement: Ask: How is the candy we made different from each of the ingredients we used in making the candy? (Change - shape, degree of wetness or dryness, texture, color, etc.)

Define Matter Scoring Rubric

4 points	correct, complete, detailed
3 points	partially correct, complete, detailed
2 points	partially correct, partially complete, lacks some detail
1 point	incorrect or incomplete, needs assistance

Scoring Criteria	4	3	2	1
Followed directions and completed activity.				
Drew sequence of events in proper order.				
Colored in face to describe candy.				
Used words or drawings to describe how the candy smells.				
Used words or drawings to describe how the candy looks.				
Used words or drawings to describe how the candy tastes.				
Used words or drawings to describe how the candy feels.				

Total Points_____

What happens to candies in water? Students conduct systematic observations to discover what kind of color patterns result when candies are placed in water.

Materials: white 6 inch plastic desert plates (available in picnic/party supplies area of grocery, discount, or hobby stores), pitcher of room temperature water, hard candy with colorful coating such as M & M's®, Gobstoppers®, assorted hard candies, etc. Note: M & M's have the advantage of coming in five colors, dispersing only one color, and having a flat shape making them easier for young children to put in position on a plate without rolling out of position. The Gobstoppers® produce spectacular patterns in water but have the disadvantage of going through several color changes and having a round shape which may easily roll out of position. We have tried assorted hard candies and they will produce patterns in water but the colors will not be as bright.

Teaching Tips: The **Observing Matter** activity should be completed before engaging students in this investigation. Ask students to recall what happened when they observed a Gobstopper® in water. Hold up a white 6 inch plastic plate and elicit suggestions for ways that candies could be placed on the plate to form color patterns in the water. Ask students to predict what pattern the color in the candy might produce in the water. After students have drawn in their pattern and prediction, have them place the candies on a dry plate. Fill a pitcher with room temperature water and pour water in each plate so that it does not go over the rim of the plate (about 3/4 full of water). Have students draw the patterns that result. Then have students describe the shapes and colors produced by the interaction of the candies with water. Ask them to draw a picture of another way to arrange the candies to produce color patterns in the water. Note: The following diagrams show the patterns resulting from possible arrangements of the candies:

Connect to Content

Inquiry/Properties of Objects and Materials: Science is a dynamic process in which scientists search for the "best" answer to questions about the universe and the way it works. Teachers must help students to think of science as a quest, not as an acquisition; and as an ongoing venture, not as a finished product.

Full inquiry involves asking a simple question, completing an investigation, answering the question, and presenting the results to others. Primary students begin to develop the physical and intellectual abilities of scientific inquiry:
• Ask a question about objects, organisms, and events in the environment.
• Plan and conduct a simple investigation.
• Employ simple equipment and tools to gather data and extend the senses.
• Use data to construct a reasonable explanation.
• Communicate investigations and explanations.

Gobstoppers®: The pigments used to color the candies are dispersed by the water. The color patterns produced by placing the candies in different positions are the result of the interaction of water molecules randomly moving in all directions dispersing the candy pigments. When the moving water molecules encounter a barrier such as the sides of the plate, the molecules bounce off the sides and continue moving away from the barrier and the dispersion of color follows the barrier shape (curved sides of plate). When no barrier exists such as in the center of the plate, the color dispersion moves through the water equally in all directions until a barrier such as large pigment molecules stop the direction of movement.

Constancy, Change, and Measurement:

Ask students to explain why they think the patterns in the water are different when you place the candies in different positions in the water. (Change - help students focus on cause and effect in constructing explanations, i.e., how many candies are on the plate and what is the resulting pattern produced in the water or where are the candies located on the plate and what is the resulting pattern)

Investigate Matter Scoring Rubric

4 points   correct, complete, detailed
3 points   partially correct, complete, detailed
2 points   partially correct, partially complete, lacks some detail
1 point   incorrect or incomplete, needs assistance

Scoring Criteria	4	3	2	1
Followed directions and completed activity.				
Recorded predicted pattern in water.				
Observed and recorded resulting pattern in water.				
Described pattern using words and/or drawings.				
Planned a new pattern to investigate.				

Total Points   _____

Observe Matter

1. Observe your candy.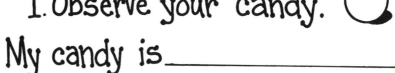

My candy is _____

_____

2. Watch the candy change in water.
3. Name and color the changes.

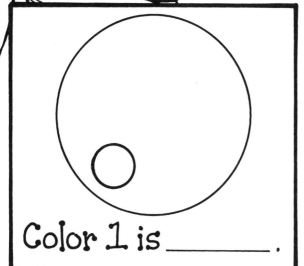

Color 1 is _____.

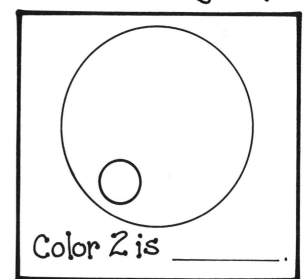

Color 2 is _____.

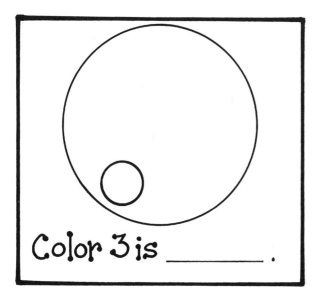

Color 3 is _____.

Color 4 is _____.

Communicate Matter

1. Color the graph to show the number of candies.

	red	orange	yellow	green	purple
10					
9					
8					
7					
6					
5					
4					
3					
2					
1					

2. Which color has the most ?_____

Which color has the least ?_____

Which colors have the same?_____

3. Write one sentence about the graph.

Classify Matter

1. Sort your candies into groups.
2. Name each group.

Measure Matter

1. Observe different candies.
2. Count out the number of candies you think will fill the container.

3. Cover the container with the Mystery Tube. Put in the counted candies.

4. Lift the Mystery Tube.

5. How many candies fell out?

6. How many more candies would make it full?

7. How many candies does the container hold when full?

candy 1

candy 2

candy 3

Matter

Predict Matter

1. Shake your box of candy.

2. Color in the first circle to show the color candy you see.

3. Shake and color 10 times.

○ ○ ○ ○ ○ ○ ○ ○ ○ ○
1   2   3   4   5   6   7   8   9   10

4. Write the number that tells how many candies you predict are in the box.

5. Count the candies and write the real count.

	Prediction	Real Count
○ red		
○ orange		
○ yellow		
○ green		
○ purple		

Infer with Matter

1. Choose a candy.

2. Observe but do not taste.

Draw and tell about your candy.

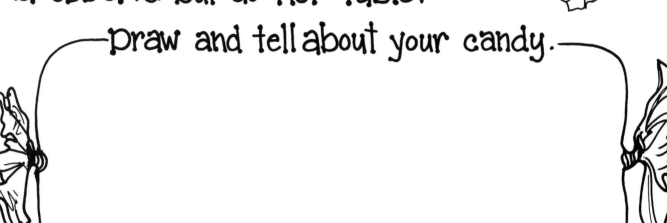

3. Infer the flavor. What do you think it will taste like? _____

Why? _____

_____

4. Taste the candy. Describe the flavor.

_____

5. What surprised you?

_____

Make a Model of Matter

Invent a candy and wrapper.

1. What ingredients will you use?

2. Make your candy. Design a wrapper.

3. Draw your candy.

4. Draw your wrapper.

What is Matter?

Define Operationally.

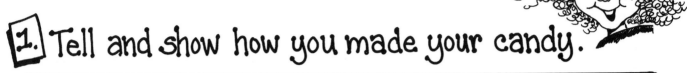

1. Tell and show how you made your candy.

Step 1	Step 2
Step 3	Step 4

2. How is your candy?

Great   Good   OK   Poor

3. Describe your candy.

Smell	Look
Taste	Feel

Matter

Investigate Matter

What happens to candies in water?

Prediction

1. Pick colored candies.

2. Put the candies on a plate. Draw your pattern.

3. Draw what you think will happen in water.

4. Add water. Wait and watch. 👀

5. Draw what you see. 👀

6. Tell about the pattern.

Results

7. Plan a new pattern.

Matter

DINOSAURS and FOSSILS

National Education Standards
Unifying Concepts and Processes: Evidence, Models, and Explanation

Properties of Earth Materials	Fossils provide evidence about the plants and animals that lived long ago and the nature of the environment at that time.
Science as a human endeavor	Although people using scientific inquiry have learned much about the objects, events, and phenomena in nature, much more remains to be understood. Science will never be finished.
Abilities Necessary to do Scientific Inquiry	As students focus on the processes of inquiry, they develop the ability to ask scientific questions, investigate aspects of the world around them, and use their observations to construct reasonable explanations for the questions posed.
Understanding About Scientific Inquiry	Scientists develop explanations using observations (evidence) and what they already know about the world (scientific knowledge). Good explanations are based on evidence from investigations.

Evidence, Models, and Explanation: Evidence consists of observations and data, which are used to construct scientific explanations. Physical or conceptual models that correspond to real objects and events can be used to explain and understand things and how they work. Evidence from fossils indicates that dinosaurs were most likely reptiles since they had skeletons like those of reptiles today. Fossils also indicate that dinosaurs had leathery skin that protected them from losing moisture through the surface of the body and drying up in the sun. Like most reptiles today and as far as anyone knows, all dinosaurs laid eggs. It was once thought that dinosaurs buried their eggs in the ground and the baby dinosaurs hatched and were on their own. However, discoveries of duckbilled dinosaur nests in Montana and Asia suggest that some dinosaurs guarded and cared for their young. A baby reptile was probably protected by a hard- or leathery-shelled egg until it developed inside the shell and was ready to hatch.

Science as a Human Endeavor: Beginning in grades K-4, you can stimulate students' natural inclination to ask questions and investigate their world. Primary students can conduct investigations that begin with a question and progress toward communicating an answer to the question. Encourage students to investigate and think about explanations for their observations. Note: It was logical to classify dinosaurs, along with lizards, as reptiles since dinosaur means "terrible lizard." Sir Richard Owen proposed the name dinosaur for a new fossil found in 1841. However, much of the evidence he submitted suggested the fossil belonged to an animal that was nonreptilian and may have been more like a mammal. Two of the nonreptilian features he described are massive (rather than long and skinny) body size and body elevated above the ground. Evidence for even more nonreptilian features has been found since then, including the ability to sustain speed, mammal-like predator-to-prey ratios, possible warm bloodedness, and mammalian bone structure.

Observe Dinosaurs

Materials: for each student: 1 hollow plastic egg, 1 plastic dinosaur that will fit inside the egg, color crayons

Teaching Tips: Hollow plastic eggs are abundantly available from discount and hobby stores in the months of March and April. Plastic dinosaur models can be found year round at discount stores usually in a large display area in the toy department that includes packages of plastic animals and other models.

Fill each egg with a model of a dinosaur. Hide the eggs around the classroom. Gather your students in a circle and tell them that they will go on an egg hunt around the room. Direct students to find an egg, NOT to open the egg, and return to the circle and sit down. When all students have found an egg (you may have to help some of them), tell students that at the count of 3 they can open their egg to find a model of something that hatches from an egg. Count 1 . . . 2 . . . 3 . . . open your egg! Explain that dinosaurs are known to lave laid eggs and that some fossil eggs have been found with tiny dinosaurs inside them.

Point out that all dinosaurs lived on land. Other reptiles, such as plesiosaurs, swam in the seas, and pterosaurs, had leathery wings and flew in the skies. However, no dinosaur ever swam or flew. Dinosaurs became extinct 65 million years ago. Ask student to observe their dinosaur carefully. Create a word wall by having students describe their dinosaur and writing down the descriptive words with an icon so students can identify each word. Then show the students the recording sheet and tell them to draw a side view and a front view of their dinosaur. Tell students that dinosaurs are extinct so we don't really know what color they might have been; however, we do have impressions of their skin in mummified dinosaurs which show that they had leathery skin. This protected dinosaurs from losing moisture through the surface of the body and drying up in the sun. Tell students that dinosaurs belong to the group of animals known as reptiles. Help students name reptiles and make a list of each type: snakes, turtles, lizards, alligators, and crocodiles. Have students describe the color of some of the reptiles listed. Encourage them to color their drawings to show what they think the colors might have been based on the color of other reptiles.

Make a Maiasaurus (good mother lizard) nest in sand on the playground. The nest should be about 3 1/2 feet high and 6 feet in diameter and contain 15 to 25 oval grapefruit-sized eggs (balloons) arranged in a circle. Explain to students that this is a model of a dinosaur nest of eggs.

Inquiry/Properties of Earth Materials/Dinosaurs: The remains of a nursery where duckbilled dinosaurs laid their eggs and young developed in safety was found in Montana. The skeletons of an adult, several youngsters, and some hatchlings were found in a fossilized nest. The remains of several other dinosaur nests with eggs were found nearby. The nests were made of mounds of mud, which had become solid rock and measured 10 feet across by 5 feet high. The nests were about 23 feet apart, which would have left room for a mother duckbilled dinosaur to lie beside her eggs.

A mother the size of a duckbilled dinosaur would be too heavy to sit on her eggs. The eggs were laid in a hollowed-out spot in the middle of the nest mound. The eggs were arranged in circles, layer on layer, in the fossilized nests. The mother probably covered each layer with soil or sand until she was finished laying her eggs. Then she probably covered the whole nest with soil or sand to keep the eggs warm while they incubated and keep them hidden from predators.

The egg of a duckbill dinosaur was about 7 inches long (about 3 times the size of a chicken' egg) and had a tough waterproof shell to protect the developing baby inside. The newly hatched dinosaurs were about 14 inches long and their mother probably brought them food until they were able to find food on their own. Since duckbilled dinosaurs nested in groups, there would have been some adults around at all times to guard the young while other adults gathered food. Once the young were strong enough, they could move in the herd where they had the safety and protection of the adult dinosaurs. Living in a herd had advantages for dinosaurs such as warning one another of predators and other dangers. However, at the breeding season, rival males probably would have taken part in fierce head-butting battles to win mates and determine the leader of the herd. The duckbill dinosaurs probably came back to the same nest site year after year.

Evidence, Models, and Explanation: Ask: How is your model of a dinosaur like a real dinosaur? (Same shape) How is your model dinosaur different from a real dinosaur? (Different size, color, materials, etc.) What can a dinosaur model help us understand about real dinosaurs? (Models help us observe their features/characteristics)

Observe Dinosaurs Scoring Rubric

4 points   correct, complete, detailed
3 points   partially correct, complete, detailed
2 points   partially correct, partially complete, lacks some detail
1 point   incorrect or incomplete, needs assistance

Scoring Criteria	4	3	2	1
Followed directions and completed activity.				
Drew and colored side view of dinosaur.				
Drew and colored front view of dinosaur.				
Made a word list.				
Competed two sentences about their dinosaur using the word list.				

Total Points _____

Classify Dinosaurs

Materials: plastic model dinosaurs or cards with pictures of dinosaurs, yarn loops, color crayons

Teaching Tips: Have students sit in a circle. Pass out a dinosaur (card or plastic model) to each student. Go around the circle and have each student describe his/her dinosaur. Demonstrate how to sort the dinosaurs into groups by putting a large yarn loop on the floor and asking students who have dinosaurs that walked on four legs to put their dinosaurs in the loop. Then put down another yarn loop and ask students who have dinosaurs that walked on two legs to put their dinosaurs in the loop. Ask students for other "rules" that could be used to sort the dinosaurs into groups. Make a list of the "rules" on chart paper or on the chalkboard. Possible rules include: size (i.e., smaller or larger than a chicken), having or NOT having a type of attack or defense structure (jaws, teeth, whip-like tails, tail spikes or clubs, claws, head horns or spikes, armored plates of bone), plant or meat eaters, long or short necks, etc. Encourage and accept student generated labels such as "gentle giants" for large plant-eating dinosaurs, "terrible lizards" for meat-eating dinosaurs, "armored dinosaurs" for those with armor, horns, plates, and frills, duckbilled dinosaurs for large plant-eaters with mouths like ducks, and "other prehistoric creatures" for things that swam or flew. Hand out the student activity sheets and have students work collaboratively to sort and classify a set of dinosaurs. Have students look at each other's sorts and guess the "rule" used to sort the dinosaurs. Which sorts were easy to guess? Which sorts were difficult to guess?

Connect to Content

Inquiry/Properties of Earth Materials/Dinosaurs: More than 300 different kinds of dinosaurs have been discovered since 1841. Paleontologists have divided the dinosaurs into two separate orders of reptiles: lizard-hipped dinosaurs (saurischians) and bird-hipped dinosaurs (ornithischians). The lizard-hipped dinosaurs had hips (pelvis bones) shaped like lizards but they did not walk like lizards. About 55% of dinosaurs were saurischians and about 45% were ornithischians. The bird-hipped dinosaurs had hips that were similar to those of birds. Most bird-hipped dinosaurs were plant eaters. They had an extra set of toothless bones at the front of the skull, which were covered by a beak made of horn for cropping leaves. Many ornithischians, like Stegosaurus and Triceratops, had strange horns, frills, spikes, or plates on their bodies. Both the lizard-hipped and bird-hipped dinosaurs included dinosaurs that walked on two legs and others that walked on all fours. The lizard-hipped dinosaurs were more varied in their lifestyles than the bird-hipped dinosaurs and included fierce, flesh-eating predators as well as plant eaters. Both groups included dinosaurs that were less than 3 feet long. But the largest lizard-hipped dinosaurs, the sauropods, were enormous (at least 98 feet long) while the largest bird-hipped dinosaurs, the duckbilled dinosaurs, were only 42 feet long and built more lightly.

One of the secrets to the successful existence of the dinosaurs was that they evolved a better way of moving than other reptiles. Most early reptiles moved like lizards with their legs sprawled out sideways. However, dinosaurs' legs went straight down underneath their bodies. This meant their legs could carry more weight, which enabled dinosaurs to become larger and heavier. By lengthening their straight limbs, dinosaurs could take longer strides to move farther and faster. The proportions of the legs were dependent on the size of the dinosaur and whether it moved on two legs or four legs.

Evidence, Models, and Explanation: Ask: What new information did you learn about dinosaurs by looking at the paper or plastic model dinosaurs? (How they walked, what characteristics or features they had, what defense or attack mechanisms they had, etc.)

Classify Dinosaurs Scoring Rubric

4 points   correct, complete, detailed
3 points   partially correct, complete, detailed
2 points   partially correct, partially complete, lacks some detail
1 point   incorrect or incomplete, needs assistance

Scoring Criteria	4	3	2	1
Followed directions and completed activity.				
Dinosaurs were sorted into two groups based on common features or characteristics.				
Feature/characteristic used for sorting was identified.				
Observed and analyzed patterns to determine the sorting scheme used by other groups.				

Total Points_____

Communicate About Dinosaurs

Materials: plastic model dinosaurs or cards with pictures of dinosaurs, graph (see directions below), scissors, color crayons, 18" X 12" construction paper for individual graphs

Teaching Tips: Make a graph six feet long and two feet wide on a sheet of butcher paper or a white shower curtain. Use a meter stick to draw a grid with four inch squares (6 X 18). Make two labels that say "Walked on 2 Legs" and "Walked on 4 Legs."

Have students sit in a circle. Pass out a dinosaur (card or plastic model) to each student. Go around the circle and have each student describe his or her dinosaur. Demonstrate how to graph the dinosaurs into groups by putting a large grid on the floor and asking students who have dinosaurs that walked on four legs to put their dinosaurs next to the label that says "Walked on 4 Legs." Then ask students who have dinosaurs that walked on two legs to put their dinosaurs on the grid next to the label that says, "Walked on 2 Legs." Ask questions such as, "How many more dinosaurs walked on ___ legs than ___ legs." Ask students for other ways that to graph the dinosaurs. Make a list on chart paper or on the chalkboard. Possible ways to graph dinosaurs include: size (i.e., smaller or larger than a chicken), having or NOT having a particular type of attack or defense structure (jaws, teeth, whip-like tails, tail spikes or clubs, claws, head horns or spikes, armored plates of bone), plant or meat eaters, long or short necks, etc. Graph the dinosaurs another way. Ask questions such as, "Which group has the most dinosaurs? Which group has the fewest dinosaurs? How many fewer dinosaurs are ___ than are ___." Hand out the student activity sheet and have students work collaboratively to cut out and graph a set of dinosaurs. After students have created a graph, ask them to glue it to a piece of 18" X 12" construction paper. Tell students to label each group on the graph. How did the graph help to organize the dinosaurs?

Connect to Content

Inquiry/Properties of Earth Materials/Dinosaur Tails: All dinosaurs had long tails whether they were 90 feet long or the size of a chicken. Many used their tail to balance the weight of the front part of the body that allowed them to run on their hind legs.

Properties of Earth Materials/Dinosaur Weapons and Armor: Many dinosaurs had horns and spikes on their heads that they may have used to charge an enemy. Marks on tailbones indicate that large muscles powered the tail, allowing dinosaurs to lash its whip-like end from side to side. Some dinosaurs had bony spikes on their tails or a bony club at the end of their tails to defend themselves. The plates of bone, which covered much of their bodies, protected armored dinosaurs. Sharp claws were probably used to seize prey. Massive jaws and dagger-like teeth could be used to kill and rip apart prey.

Evidence, Models, and Explanation: Ask: How did the graph help you analyze the features of the model dinosaurs? (it was easier to tell and count how many dinosaurs were in each group) Was there anything about the dinosaurs that you couldn't find out by looking at the paper dinosaurs? (size, color, texture of skin, etc.)

Communicate About Dinosaurs Scoring Rubric

4 points   correct, complete, detailed
3 points   partially correct, complete, detailed
2 points   partially correct, partially complete, lacks some detail
1 point   Incorrect or incomplete, needs assistance

Scoring Criteria	4	3	2	1
Followed directions and completed activity.				
Observed features of dinosaurs and suggested ways they are different.				
Cut out and glued the cards down to create a graph.				
Labeled each group of dinosaurs on the graph.				

Total Points_____

Predict About Fossils

Materials: student recording sheet, scissors, color crayons, and tape
Demonstration: plastic shoe box half filled with moist sand, three plastic dinosaur models (plant-eater, meat-eater, scavenger), model of dinosaur bones, flat piece of sandstone or shale

Teaching Tips: Assess prior knowledge by asking how something might become a fossil. Explain that most animals live and die without leaving any evidence of their existence such as a fossil. Conditions have to be just right for an animal to become a fossil. You may want to act this out using plastic dinosaurs and a plastic shoebox half filled with moist sand. Tell students to imagine that millions of years ago, a dinosaur came down to a river to drink but was killed there by a meat-eating dinosaur (show two dinosaurs - use discretion about how graphic you want to make the demonstration). After this predator ate its fill, most of the soft body parts were eaten by scavengers and the rest slowly rotted away (remove dinosaur and replace with dinosaur skeleton). Then the bones of the skeleton sank into the sand and mud of the riverbank (push dinosaur bones into sand). Explain that this is how the fossilization process begins. As time went on, whenever the river flooded, sand and mud was laid over the skeleton until it was buried deeply by sand and mud (put sand over the bones). The river water contained minerals which filtered down and cemented the particles of sand and mud together turning the sand and mud into hard rock (have a flat piece of sandstone or shale to show). The minerals in the water also slowly replaced the bones and filled up spaces where the brain and spinal cord had been. Explain that the skeleton kept its original shape but was changed into rock. Over time, more sand and mud were deposited on top of the skeleton that had changed into rock. Later, forces deep inside the earth pushed up the land and then the rocks were eroded by rain and wind until the skeleton once again was near the surface of the ground. Then a scientist called a paleontologist recognized the bones and discovered the fossil of the dinosaur. Ask students to predict the sequence of events, which changes an animal into a fossil by coloring, cutting out, and then taping the diagrams on the activity sheet in order. This is often referred to as a sequence chain and will help students develop an awareness of the difficulty of collecting fossils and the sequential process of fossilization.

Connect to Content

Inquiry/Properties of Earth Materials/Fossils:
Event 1: After a dinosaur dies, the soft parts of its body are eaten by other animals and decays. **(Dies)**
Event 2: The dinosaur's skeleton sinks into the mud by the river. **(Sinks)**
Event 3: Over millions of years, more layers of mud are deposited over the skeleton. **(Covered)**
Event 4: The land rises and erosion removes some of the rock above the skeleton. **(Uncovered)**
Event 5: The last layers of rock covering the skeleton are eroded to reveal the fossil bones. **(Discovered)**
Paleontologists, scientists who study life of the past, go on fossil-hunting expeditions. If they are searching for the remains of Jurassic dinosaurs, they will travel to a place where rocks deposited during the Jurassic period are on the surface of the Earth. Erosion may reveal fossil bones that have been buried for millions of years. Fossil hunting is easiest in deserts were plants are not covering the eroded rocks. Once a skeleton is found, the rock above it must be carefully removed and the bones are covered first with wet tissue paper and then with a layer of sacking which has been soaked in plaster of Paris. After the plaster has hardened, the rock beneath the bones is cut away. The entire section is then turned upside down and more plaster is applied to make a covering around the bones.

Evidence, Models, and Explanation: Ask: How did the demonstration help provide an explanation of how dinosaurs and other things become fossils? (The demonstration modeled a process that takes place over a long period of time and will repeat in the future to create more fossils)

Predict about Fossils Scoring Rubric

4 points   correct, complete, detailed
3 points   partially correct, complete, detailed
2 points   partially correct, partially complete, lacks some detail
1 point   incorrect or incomplete, needs assistance

Scoring Criteria	4	3	2	1
Followed directions and completed activity.				
Observed the demonstration.				
Cut out pictures depicting the sequence of events in forming fossils.				
Correctly sequenced and taped the cards together in the order of events which form fossils.				
Verbally described the sequence of events in fossil formation.				

Total Points_____

Measure Dinosaurs

Materials:
For each student: 4"-5" of butcher paper, color crayons, scissors
Use markers to make a drawing of one or more of the following dinosaurs on an old sheet or butcher paper.

Teaching Tips: Select one or more of the dinosaur card drawings and make a transparency of it. Move the overhead projector until the size projected matches the size listed on the dinosaur chart. Once the size is correct, trace the dinosaur outline on an old sheet or butcher paper. The dinosaurs Coelophysis, Deinonychus, Ornithlestes and Pachycephalosaurus will fit on a double bed size sheet. A sponge and fabric paint (if using a sheet) or tempera paint (if using butcher paper) can be used to fill in the dinosaur outline with textured color. Larger dinosaurs can be drawn full size on the blacktop. If other teachers at your school are doing these activities, each teacher can make one of the dinosaurs and all the teachers can share the drawings by hanging them in a common area such as the hallway.

      Give each student a piece of butcher paper as long as the student is tall. Have students work in pairs to trace their partner's body outline on the butcher paper. Have students observe one of the dinosaurs. Ask questions such as "How many students would it take to be as tall as this dinosaur?" Have a students place their body outlines end to end up the height of one of the larger dinosaurs and count the body outlines together. Ask students to estimate how many body outlines long the dinosaur is before measuring the length with body outlines. Then have them place their body outline along the length of one of the dinosaurs and count the body outline together. Have students record the "body measurements" on the measuring data sheet. You may want to introduce the idea of "rounding off" by asking students what they should do if the dinosaur height or length doesn't come out to exactly the end of a body outline. Come up with a rule such as, "If the measurement is more than half of a body outline, then we will count it and "round up" but if it is less than half, then we will "round down" and won't count it. Make a drawing of Comsognathus using the same transparency technique described above. Have students compare the size of this dinosaur to one or more of the large dinosaurs. Then challenge students to think of other objects that could be used to measure the dinosaurs. Ask student to measure and record their measurements on their activity sheet.

Connect to Content

Inquiry/Properties of Earth Materials/Dinosaurs: Many students think that dinosaurs were big but the sizes ranged from the 85 feet long Ultrasaurus down to the chicken-sized Compsognathus.

Apatosaurus: Length: 21 meters (70 feet) from tail to nose, Shoulder height: 4.5 meters (15 feet), Neck: 6 meters (20 feet) long

Tyrannosaurus: Length: 13 meters (43 feet), Height: 6 meters (18.5 feet) Head: 1.2 meters (4 feet) long, Jaws: 90-cm (3 feet) longs, Talons: 20-cm (8 inches) longs, Arms: 30 inches (76 cm) long

Ultrasaurus: Length: 26 meters (85 feet), Height: 14 meters (45 feet), Front legs: 5 meters (17 feet) long, Neck: 6 meters (20 feet) long, Shoulder height: 7.5 meters (25 feet)

Compsognathus: Length 60 cm (2 feet), Height: slightly over 30 cm (1 foot)

Evidence, Models, and Explanation: Ask: How did the life-sized model of the dinosaur compare to the life-sized model of your body? (Larger or smaller depending on the dinosaur that was measured, different shape,etc.)

Measure Dinosaurs Scoring Rubric

4 points   correct, complete, detailed
3 points   partially correct, complete, detailed
2 points   partially correct, partially complete, lacks some detail
1 point   incorrect or incomplete, needs assistance

Scoring Criteria	4	3	2	1
Followed directions and completed activity.				
Made a paper outline of body.				
Measured and recorded measurements of dinosaurs using body outline and rounded off to nearest body unit.				
Measured and recorded measurements of dinosaurs using other non-standard or standard units.				

Total Points_____

Materials: 21 bone pieces, scissors, glue, 18" X 24" sheet of newsprint

Teaching Tips: Have groups of students cut out the 21 bone pieces. You may want to use pairs or groups of four depending on the social skills of your students. Have the group divide the pieces among the members of the group. There will be one piece left over when the pieces are divided between two people (10 pieces each) or among four people (5 pieces each). Instruct students to have one person in the group place the extra bone on the sheet of newsprint. Then have students take turns placing bones next to the bones that are already on the newsprint until all the bones have been placed on the paper. The next step is to ask the groups to work together to rearrange the bones until the groups are in agreement about the reconstruction of the fossil skeleton. Tell groups to think of a name for the animal and infer how the animal moved and what it might have eaten. Have groups take turns displaying their fossil reconstruction, telling the name that was given to the creature, and explaining how it might have moved and what it might have eaten.

Ask students to imagine looking up to see a fur-covered creature gliding through the air on wings longer than the wings of a small airplane. Explain that this prehistoric creature is not a dinosaur but a pterosaurs (TER a SORZ) which means "fling lizard" and might have been a common sight millions of years ago during ancient times called the Triassic and Cretaceous. The largest pterodactyl, Quetzalcoatlus had a wingspan of 45 feet. It probably soared from cliffs on its graceful sail-like wings. These giant pterosaurs shared the skies with many smaller-winged pterosaurs such as the one they have just reconstructed, *Scaphognathus crassiorostris* (ska FOG NA THAS KRAS i ROS tris).

Explain that this prehistoric creature was found in Germany in 1826 by the German Scientist, August Goldfuss. It was named *Scaphognathus crassiorostris* (ska FOG NA THAS KRAS i ROS tris), and the bones are about the same size as the fossil bones found in Germany. This pterosaur was about the size of a large bat. Show students the fossil cast of *S. Crassirostris* and point out the long, rounded jaw and beak of this pterosaur and explain that the animal's name is actually a description of its distinctive head. *Scaphognathus* means "boat-shaped jaw" and *Crassirostris* means "large beak."

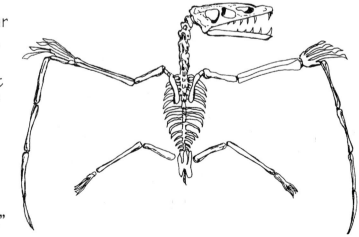

The illustration is a picture of the fossilized bones of the S. Crassirostris shown as they were found in a limestone quarry. Most of the bones are not broken but a few bones have changed orientation in the skeleton since the animal's death. For example, the rib has moved away from the animal's ribcage to become fossilized underneath the head.

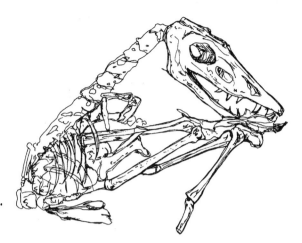

Have each student draw a picture of S. Crassirostris in its environment. Ask students what might have been in its environment that it could eat. (Fish) Where would the S. Crassirostris have to go to get fish? (Water) Tell students to consider this when drawing the creature in its environment.

Connect to Content

Inquiry/Properties of Earth Materials/Fossils: There were two kinds of flying reptiles: rhamphorhynchs and pterodactyls. All had wings attached to the enormously elongated fourth finger of each hand. The other three fingers were short and tipped with sharp claws. The wings were also attached to the sides of the pterosaur's body, possibly at hip level. Rhamphorhynchs were the earliest and most primitive pterosaurs. They had short legs and a long bony tail, which made up about half the animal's length. This tail would have been held straight out during flight and may have helped to balance the front part of the body. Most rhamphorhynchs had large heads and jaws filled with sharp teeth. Pterodactyls developed during the Jurassic. They had the same general structure as the rhamphorhynchs, but the tail was shorter and the neck and skull longer. No one knows how pterosaurs moved on the ground. Some experts think that they crawled along using the claws on their wings as well as their feet.

The skeleton of S. Crassirostris was found in limestone that formed in a warm seawater lagoon. Pterosaurs may have fed in a variety of ways. Some of the smaller, fast-flying pterosaurs probably chased prey in mid-air, like today's insect-eating birds. Others may have snatched fish from the water in long, narrow jaws. The wings were made of a delicate flap of skin. If this flap of skin tore, it probably could not fly. This might explain how S. Crassirostris might have become extinct. However, recent research shows that wings may have been strengthened by tough fibers. Scientists believe that pterosaurs did actually flap their wings, rather than just gliding through the air. The earliest know bird, Archaeopteryx lived 150 million years ago and had features of both reptiles and birds. Like reptiles, it had toothed jaws and a long, bony tail. Like birds, it had feathers and wings.

Step-by-step directions for assembling the bones: The drawings of the bones of S. Crassirostris are about the same size as the fossil bones found in Germany. Using the illustration as a reference, make a model of the skeleton of S. crassirostris from the bones. Find the two bones that make up the long, curved neck. Then locate the pieces that make up the backbones and ribs. Identify the shoulder blades and pelvic bones. Arrange these pieces to make the trunk. Note that the legs were short compared to the arms. Unlike other vertebrates, the upper and lower parts of the leg were each composed of two long, fused bones. The hands were much larger than the feet. S. Crassirostris had five fingers on each hand but when the pterosaur was in flight, four of the fingers protruded from the middle upper edge of the wing. The remaining set of long finger bones extended from the four protruding fingers to the wing tip. The bones of this fifth finger were the main support for the lower extension of the wing. Find the two pieces that make up the hand and note how the arm bones and hand bones fit together to make up the wings. Arrange the limb bones off the trunk. Then glue the bones of the skeleton model of S. crassirostris in position on a sheet of 18" X 24" newsprint.

Evidence, Models, and Explanation: Ask: When putting the bones together, what other animals did you think about (answers will vary but may include birds, humans, etc.) How did information about other skeletons help you make a model of S. crassirostris? (You could infer from other skeletons how this skeleton might be constructed)

Inferring about Fossils Scoring Rubric

4 points   correct, complete, detailed
3 points   partially correct, complete, detailed
2 points   partially correct, partially complete, lacks some detail
1 point   incorrect or incomplete, needs assistance

Scoring Criteria	4	3	2	1
Followed directions and completed activity.				
Cut out the bones.				
Worked collaboratively to reconstruct the skeleton and glue down the bones.				
Drew a picture of the creature in its probable environment.				
Describe what the creature probably ate and how it moved.				

Total Points_____

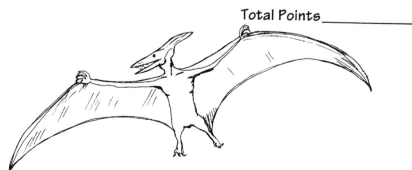

Making Models

Materials:
Aluminum foil, pipe cleaners, clay, assorted pasta, assorted scrap materials

Teaching Tips: Directions: Have students compare the bones of the two skeletons with the bones in their own bodies. Ask students to walk like a dinosaur. Have them get on the floor in a crawling position. Then have them place their arms directly under their shoulders, with their hands arched up so only their fingers touch the floor. Tell the students that their palms should not be touching the floor. Then have students place as much weight as possible on their arms to see how it feels. Challenge students to hold his position while you rapidly count to 10. Ask students to crawl in this position. Say that they are now walking like dinosaurs.

Then have students walk like a lizard. Have them keep their body and feet in the same position and hold their upper arms and elbows out to their sides (as in the lizard-like skeleton), with their palms flat and hands turned out. Ask students to shift their weight forward on their arms. Then have them crawl forward in the sprawling lizard-like position. Say that they are now walking like lizards.

Ask students to describe the differences in walking like a dinosaur and walking like a lizard. Guide the discussion to verify that crawling like a dinosaur was much easier, faster, and more comfortable. Moving like a lizard was extremely awkward. Students may have observed that when they walked like a lizard, the bodies had to wriggle and twist in the sprawling position like lizards do when they move.

After students have had an opportunity to role-play dinosaurs, have them make aluminum foil dinosaurs, pipe cleaner dinosaurs, clay dinosaurs, pasta-saurs, or dinosaur puppets. Aluminum dinosaurs can be made by ripping off three sheets of aluminum foil that are 4 feet long each. Make an X with two of the sheets and then place the third sheet across the center of the X. Crumple the center part (X) of the aluminum foil strips together just enough so that the pieces of foil stick together. Make the head and tail of the dinosaur by loosely crumpling each end of the centerpiece of aluminum foil into a long, thin shape. Start at the end of the remaining pieces of foil to make the legs by crumpling them toward the body. To complete the body of the dinosaur, finish scrunching together the center section of the foil. Finish the dinosaur by squeezing the foil together and molding it into the shape of the dinosaur. Pasta-saurs can be made by gluing pasta on construction paper to make a dinosaur skeleton. Pipe cleaners, clay, and other materials can be used to make dinosaur models - the possibilities are limitless. Have students list the materials they used in their dinosaur model, draw the model, and fill out the "Mini-Museum Card." Then ask students to tell how their model is like a dinosaur and how it is different from a dinosaur.

Connect to Content

Inquiry/Properties of Earth Materials/Prehistoric Creatures: Dinosaurs are members of a unique group of reptiles known as archosaurs which also includes crocodiles, flying reptiles (pterosaurs), and possibly birds. They ranged in size from a small bird to a small building. Dinosaurs lived during a very specific time period, had special skeletal structures, and lived in certain environments. Dinosaurs lived during the Mesozoic era (230-66 million years ago). This era is known as the Age of The Dinosaurs; however, dinosaurs from the earliest Mesozoic (Early or Middle Triassic period, 245 million years ago) have not been discovered. Many animals listed as dinosaurs are not only structurally different from dinosaurs but also lived before or after the dinosaurs. Dimetrodon lived during the Permian period of the Paleozoic era (286-245 million years ago) so it is not a dinosaur. The woolly mammoth may be "big, scary, and old" but it lived only tens of thousands of years ago, much younger than dinosaurs. Dinosaurs vanished from the earth more than 60 million years before the earliest humans existed. Dinosaurs lived on land; however, they were able to spend time in water like we swim or play in the water. Animals that were principally fliers such as pteranodons or lived in the ocean such as mosasaurs, plesiosaurs, and icthyosaurs) were not dinosaurs.

Dinosaurs had special adaptations in their hip and leg structures that allowed them to support their bodies from below and move quickly (other reptiles have legs that extend from their sides. All dinosaurs belong to two orders of reptiles distinguished only on the basis of their hip structures (Ornithischia or Saurischia). A dinosaur walks on its toes like a ballet dancer, its legs are positioned directly below its body, and the body is supported high off the ground. However, when a lizard walks, its feet are planted flat on the ground, its arms and legs are off to the sides, and its body is suspended low between its shoulders and hips.

Evidence, Models, and Explanation: Ask: How is your model like a dinosaur? (Same shape) How is your model different from a dinosaur? (Different size, materials, etc.)

Make a Model of Dinosaurs Scoring Rubric

4 points   correct, complete, detailed
3 points   partially correct, complete, detailed
2 points   partially correct, partially complete, lacks some detail
1 point   incorrect or incomplete, needs assistance

Scoring Criteria	4	3	2	1
Followed directions and completed activity.				
Listed materials used in model dinosaur.				
Drew a picture of dinosaur model.				
Filled out the "Mini-Museum Card."				
Listed similarities and differences in a model dinosaur and a real dinosaur.				

Total Points_____

Materials: Paper models of stegosaurus and deinonychus, color crayons, and scissors

Teaching Tips: Cut out the paper models of stegosaurus and deinonychus and color the dinosaurs. Make sure that you do not cut the dinosaurs apart; each one it is designed to be folded so that it will be free standing. Assess prior knowledge by asking what dinosaurs ate. Then ask what kind of teeth they would need. Show pictures from library books to show the flat, blunt teeth of plant eaters and the sharp teeth of meat eaters. (A cat or fox would be good examples of meat eaters and a deer or cow would be good examples of plant eaters.) The size of a dinosaur skull and type of teeth in it reveals whether a dinosaur was a plant eater or a meat eater. If the skull has powerful jaws and long pointed teeth, it probably was a meat eater. If the skull has flat, blunt teeth for grinding or chopping, then the dinosaur was probably a plant eater. Ask students to look at your paper models of stegosaurus and deinonychus and explain what they probably ate. Then ask a student to walk like a dinosaur (on two feet or on four feet), then ask if there was any other way dinosaurs walked (on two feet or on four feet). Explain that some dinosaurs walked on two feet and other dinosaurs walked on four feet. Direct the students to look at your paper models again to determine how many feet each of the model dinosaurs walked on. Then give students copies of the paper models and ask them to cut out and color the dinosaurs being careful not to cut each one apart. Demonstrate how to fold the dinosaurs and make them stand up. Then have them observe each paper model and list the features of each one. You may want to make a word wall with your students so they can use words from the wall on their activity sheets. Than ask students to compare the dinosaurs and list the things that both dinosaurs have in common. Finally, ask students to write their definition of a dinosaur based on their observations.

Note: *Stegosaurus* was a plated dinosaur that used its turtle-like, beaked mouth to chop soft vegetation, then chopped the food with the small, weak teeth in the back of its mouth. *Stegosaurus* was thirty feet long and is the largest of the plate-backed dinosaurs. *Deinonychus* ("terrible claw") was a man-sized predator with long limbs and prodigious claws. This predator's slender legs and huge toe claws suggests a fast-running, high-kicking lifestyle. *Deinonychus* had to have a metabolism more like active, warm-blooded birds than sluggish, cold-blooded reptiles. It is argued that Deinonychus and other dinosaurs of its kind shared a recent common ancestor with birds among more primitive dinosaurs. If birds were not dinosaurs, they were more closely related to theropod dinosaurs than to any other creatures.

Inquiry/Properties of Earth Materials/Dinosaurs:
• Dinosaurs lived primarily on land.
  • Some dinosaurs walked on four legs while others walked on two legs
  • Reptiles that lived in the ocean or flew during the *Age of Dinosaurs* were <u>not</u> dinosaurs
• Dinosaurs lived between 225 and 65 million years ago.
• Dinosaur legs came straight down from its body.
  • Dinosaur means "terrible lizard" but they were <u>not</u> lizards because the legs came straight down from its body, not out of the side like a lizard
• Dinosaurs had extra openings in the skull to reduce the weight of the head & hold muscles that helped them chew.

Because this activity is designed to help students define dinosaurs; it is worth spending a little time talking about what a dinosaur is, or rather, what a dinosaur is not. If you ask students to explain what a dinosaur is, a typical response might be "an animal that is scary, big, and old." Many would also add slow, clumsy, and stupid. Some may answer with a list of well-known examples: "*Triceratops, Tyrannosaurus, Stegosaurus, and Apatosaurus.*" In fact, size is not a consideration; dinosaurs ranged in size from a small bird to a small building; scariness is not a factor either. Whereas *Tyrannosaurus* is definitely a fearsome beast, *Apatosaurus* is almost cute. Some of the student responses are certainly true, but scientists have gone a step further.

Dinosaurs are classified as members of a unique group of reptiles known as archosaurs, which also includes crocodiles, the flying reptiles (pterosaurs), other extinct animals, and possible the birds. To be a dinosaur, however, a specific type of archosaur that lived during a very specific time period had to have certain structures in its skeleton and had to live in certain environments.

Dinosaurs were unique members of the reptile class. Other reptiles, such as lizards, snakes, and crocodiles, lived with the dinosaurs, but they were not dinosaurs. Dinosaurs had special adaptations in the structure of their hips and legs that allowed them to support their bodies from below and move quickly; other reptiles have legs that extend from their sides. Mammals, fish, and amphibians also lived at the same time, but they were not dinosaurs either. It is interesting to note that in the formal classification of animals, dinosaurs are not even mentioned; all dinosaurs belong to two orders of reptiles distinguished solely on the basis of their hip structures. As a lizard walks, its feet are planted flat on the ground, its arms and legs are off to the sides, and its body suspended low between its shoulders and hips. As a dinosaur walks, it does so on its toes like a ballet dancer, its legs are positioned directly below its body, and the body is supported high off the ground.

Dinosaurs are known from only a very specific portion of earth history, the Mesozoic era (230-66 millions years ago). Although this is known as the Age of Dinosaurs, dinosaurs from the earliest Mesozoic (Early or Middle Triassic period, 245 million years ago) are not known. Many animals often considered to be dinosaurs not only are structurally different from dinosaurs but also lived before or after the dinosaurs. Dimetrodon, often found in dinosaur books, lived during the Permian period of the Paleozoic era (286-245 million years ago); it is older than the dinosaurs.

Animals like the woolly mammoth were "scary, big, and old," but they lived "only" tens of thousands of years ago, much younger than the dinosaurs. Fossil evidence indicates that human ancestors are no older than 2 or 3 million years. Clearly, the dinosaurs had vanished from the earth more than 60 million years before the earliest humans existed. People and dinosaurs did not live contemporaneously. For the sake of this discussion, birds are not considered as dinosaurs.

From kingdom to species, the grouping of animals becomes smaller and smaller based on unique characteristics. Despite their uniqueness, dinosaurs do not have their own classification. An example of the formal taxonomic classification of two dinosaurs follows:

CLASSIFICATION	TRICERATOPS	TYRANNOSAURUS
Kingdom	Animalia	Animalia
Phylum	Chordata	Chordata
Class	Reptilia	Reptilia
Subclass	Diapsida	Diapsida
Superorder	Archosauria	Archosauria
Order	Ornithischia	Saurischia
Suborder	Ceratopsia	Theropoda
Family	Ceratopsidae	Tyrannosauridae
Genus	Triceratops	Tyrannosaurus
Species	Triceratops horridus	Tyrannosaurus rex

It was only logical to classify dinosaurs, along with the lizards, as reptiles. After all, dinosaur means "terrible lizard," or reptile. The irony is that even as Sir Richard Owen was proposing the name dinosaur for the new fossil finds in 1841, much of the evidence he submitted suggested the animals were nonreptilian and may have been more like mammals. Two of the nonreptilian features he described are massive (rather than long-and-skinny) body size and body elevated above the ground. Since then, evidence for even more non-reptilian features has been found, including the ability to sustain speed, mammal-like predator-to-prey ratios, possible warm-bloodedness, and mammalian bone structure.

The saying is often heard, "If it walks like a duck, sounds like a duck, and looks like a duck, then it's a duck!" The reverse should hold true for dinosaurs: "It doesn't walk like a reptile, isn't built like a reptile, and doesn't have bones like a reptile. . .," but it's classified as a reptile. It has been suggested that dinosaurs be removed from the class *Reptilia* and given their own class, *Dinosauria* (which, incidentally, is closer to birds than to other reptiles).

Dinosaurs lived on land. They were able to spend some time in water, much as people can wade or swim in lakes and oceans, but the dinosaurs, like humans today, were creatures of the land. Animals that were primarily fliers (pteranodons) (birds are, arguably, dinosaur descendants) or lived in the ocean (mosasaurs, plesiosaurs, and ichthyosaurs) were not dinosaurs.

Dinosaur names are often chosen on the basis of these three categories: (1) place of discovery, (2) name of the discoverer or some expert in the field, or (3) description of the animal or some feature of its anatomy. The lists that follow are not intended to be all-inclusive. They merely provide examples of these three types of names.

The naming of a dinosaur after a place is very straightforward. The actual name of a site associated with the discovery is included in the name of the animal:

Albertosaurus	Alberta, Canada
Bactrosaurus	Bactria, Mongolia
Edmontosaurus	Edmonton, Alberta, Canada
Nemegtosaurus	Nemegtu, Mongolia
Zigongosaurus	Zigong, China

Infrequently, a discovery is named for a person, either a paleontologist or a benefactor of the discovery:

Lambeosaurus	for Lawrence Lambe, paleontologist with the Geological Survey of Canada
Diplodocus carnegii	for Andrew Carnegie, who financed the expedition to discover a dinosaur for the Carnegie Museum in Pittsburgh, Pennsylvania
Hadrosaurus foulkii	for William Parker Foulk, who excavated the specimen

By far the greatest number of dinosaur names are descriptive, revealing the shape, analogy to a modern animal, behavior, size, or some other anatomical feature of the animal:

Ar / rhino / cera / tops	no / nose / horned / face
Corytho / saurus	Corinthian helmet / reptile
Iguana / don	iguana / tooth
Lyco / rhinus	wolf / snout
Maia / saurus	good mother / reptile
Pro / compso / gnathus	before / pretty / jaw
Stego / saurus	roofed / reptile
Tyranno / saurus rex	tyrannical / reptile king

The names seem convoluted at first, but like so many words in the English language, they are composed of a string of Greek or Latin roots. Once those roots are known, the name of virtually any dinosaur becomes an interpretive label. The following list includes many of the roots (prefixes, suffixes, and combining forms) from which dinosaur names are created:

Greek / Latin	definition	Greek / Latin	definition
a, ar, an	no, not	mimus	mimic
acro	top	mono	single
allo	strange	morpho	shaped
alti	tall, high	mucro	pointed
angusti	sharp	nano	dwarf
apato	deceptive	nodo	lumpy
baro	heavy, pressure	nycho	clawed
bi	two	ornitho	bird
brachio	arm	pachy	thick
brachy	short	ped, pod, pes	foot
bronto	thunder	penta	five
canthus	spiked, spined	phalangia	toes
cera	horned	phobo	fearsome
coelo	hollow	placo, plateo	flat
compso	pretty	pola, poly	many
dactyl	finger	preno	sloping
deino	terrible	ptero	winged
derm	skin	quadri	four
di	two	raptor	thief
don, den	tooth	rex	king
dromaeo	running	rhino	nose
drypto	wounding	saurus	reptile, lizard
echino	spiked	segno	slow
elasmo	plated	stego	roofed
elmi	foot	steneo	narrow
gnathus	jaw	stenotes	finger
hetero	mixed	stereo	twin
lana	wooly	struthio	ostrich
lepto	slender	tarbo	alarming
lestes	robber	tetra	four
lopho	ridged	thero	beast
luro	tail	top	head, face
macro	large	tri	three
maia	good mother	tyranno	tyrant
mega	huge	velox, veloci	speedy, fast

Evidence, Models, and Explanation: Ask: How does your definition of a dinosaur reflect everything you know about dinosaurs? (Answers will vary but will include all the information students have learned doing dinosaur activities)

Dinosaurs - Define Operationally Scoring Rubric

4 points   correct, complete, detailed
3 points   partially correct, complete, detailed
2 points   partially correct, partially complete, lacks some detail
1 point   incorrect or incomplete, needs assistance

Scoring Criteria	4	3	2	1
Followed directions and completed activity.				
Listed features of Stegosaurus.				
Listed features of Deinonychus.				
Compared the dinosaurs and listed features they had in common.				
Defined dinosaurs operationally based on their observations.				

Total Points_____

Investigating - Fossil Dig

Materials: plastic shoeboxes, sand, bones or plastic dinosaurs, dividers (string & tape or rubber bands), plastic spoons, toothbrushes or inexpensive small paint brushes, egg carton for a half dozen eggs or cut so it has six compartments

Teaching Tips: You can encourage your students to assume to role of paleontologists by dressing up for this activity. If appropriate, suggest that they wear a large sun hat, jeans, and a T-shirt with rolled up sleeves. Use pictures from books illustrating a fossil dig or show the class some of the actual tools a paleontologist might use on a fossil dig: shovel, bucket, pick, spade, trowel, hammer, brush, paint brush, chisel, wheel barrow, etc. Explain that the students will do what scientists called paleontologists do to uncover fossils. Decide whether you want to do the dig as a whole class in the sandbox or as groups around plastic shoeboxes in the classroom. If you use small groups of four assign the following jobs: 2 diggers (spoons), 1 cleaner (brushes), and 1 museum curator (puts "fossil" in museum egg carton for display). Badges for *Digger, Cleaner,* and *Curator* can be cut out and laminated so that the students can wear them during the activity. The plastic shoeboxes need to be divided into six sections (see recording sheet) using string taped to the boxes or very large rubber bands stretched around the boxes. The egg cartons need to have six labeled sections that match the divided boxes (see recording sheets). Bury objects such as bones, plastic dinosaurs, or shells in some of the sections. Have each team use a spoon to carefully dig into each of the sections to unearth their "fossils." Have one student in charge of cleaning the "fossil" with a toothbrush or small paintbrush. Another student will be the museum curator and place the "fossils" in the museum (egg carton). All team members should record their "finds" by drawing or tracing each find on the Dinosaur Dig recording sheet and listing each find on the Dinosaur Dig Journal. Then have each student sequence the steps taken to dig up their fossils on the Dinosaur Dig Journal. A culminating activity is to pretend a new dinosaur has been discovered. Students give it a name, draw what it looks like, and describe its habitat. They draw a footprint and list its length, height, and weight.

Connect to Content

Inquiry/Properties of Earth Materials: Fossils Although dinosaurs have been studied for over 170 years, new species are still being discovered and we still don't understand everything about some of the dinosaurs that have been found. Dinosaurs have been found in every continent except Antarctica, where the rocks are covered by snow and ice. The great French scientist, Georges Cuvier, realized over 200 years ago that some fossil remains were so unusual that they must belong to creatures that no longer existed on Earth. Before this, many people believed that fossils were the remains of animals still living on Earth but not yet discovered. In 1822, Gideon Mantell, a British doctor, described some big teeth that his wife had found. He thought that they belonged to an extinct relative of the iguana lizard and called the fossil animal *Iguanodon*. Twenty years later, British paleontologist, Richard Owen, realized that *Iguanodon* and other huge fossil reptiles belonged to a separate group he named the dinosaurs.

Baryonyx
A very unusual dinosaur, *Baryonyx*, was discovered in Sussex, England in 1983. The dinosaur is about 28 feet long and has a very strange skull. The skull of most dinosaurs steadily increases in height from the tip of the snout to the back of the head. However, *Baryonyx* has a skill that only becomes deep from the level of the eyes backward. The snout is long and low and at the front is a group of large, outwardly pointing large teeth. This type of arrangement of teeth is not know in any other dinosaur, but is found in some fish-eating crocodiles and fossil amphibians. The teeth were used to impale fish. A big curved claw was also found with the skeleton of *Baryonyx*. The dinosaur may have used this claw to pierce fish and hold them down while it ate.

Deinocheirus
A remarkable dinosaur lived in Mongolia toward the end of the Cretaceous period. All that has been found so far by paleontologists are its enormous arms, which are nearly 8 feet long. These arms give the dinosaur its name, *Deinocheirus*, which

means 'terrible hand." The arms are slender and long and end in three fingers that bear huge, 12-inch claws. Deinocheirus could not have folded its fingers back to grasp its food. How this strange dinosaur lived will remain a mystery until a more complete skeleton is found.

Supersaurus, Ultrasaurus, and Sismosaurus.
Part of the fascination of dinosaurs is the enormous size of some of them, and no one knows for sure how big the largest of them were. In the past 20 years, the remains of larger and larger sauropods have been found that may have been over 100 feet long. They have been given names such as *Supersaurus, Ultrasaurus,* and *Sismosaurus.*

Evidence, Models, and Explanation: Ask: What evidence did we uncover to help us give an explanation of the past? (Fossils, bones, plastic dinosaurs, shells, etc.) How do these things help us understand what happened long ago? (The fossils are "clues" that can be used to help us understand what existed and happened in the past)

Investigate Dinosaurs Scoring Rubric

4 points	correct, complete, detailed
3 points	partially correct, complete, detailed
2 points	partially correct, partially complete, lacks some detail
1 point	incorrect or incomplete, needs assistance

Scoring Criteria	4	3	2	1
Followed directions and completed activity.				
Worked collaborative on the "dig."				
Predicted what might be found.				
Drew pictures of what was found in each section of the "dig" site.				
Named each item found in the "dig."				
Listed the steps taken to dig up the fossils				

Total Points_____

DINOSAUR CHART

Dinosaur Name Pronunciation	Meaning	Food	Length	Age	Fascinating Facts
Allosaurus (AL-uh-SAWR-us)	other lizard	meat	11 m (36 ft)	Late Jurrassic	The remains of more than 40 allosaurs have been collected from a single area in Utah.
Ankylosaurus (ang-kile-uh-SAWR-us)	curved lizard	plants	5 m (17 ft)	Late Cretaceaus	A blow from Ankylosaurus' club-like tail may have been powerful enough to break a predator's leg.
Apatosaurus (ah-PAT-uh-SAWR-us)	deceptive lizard	plants	21 m (70 ft)	Late Jurrassic	Apatosaurus weighed as much as five African elephants.
Brachiosaurus (BRAK-ee-uh-SAWR-us)	arm lizard	plants	22 m (75 ft)	Late Jurrassic	Brachiosaurus is one of the largest known land animals.
Coelophysis (see-lo-FISE-iss)	hollow form	meat	3 m (10 ft)	Late Triassic	Coelophysis had hollow bones as birds do today.
Compsognathus (Komp-sug-NY-thus)	pretty jaws	meat	60 cm (2 ft)	Late Jurrassic	Compsognathus was about the size of a chicken.
Corythosaurus (ko-RITH-uh-SAWR-us)	helmet lizard	meat	7 m (23 ft)	late Cretaceous	Corythosaurus was one of the duck-billed dinosaurs with an unusual crest on its head.
Deinonychus (dine-ON-ik-us)	terrible claw	meat	2.7 m (9 ft)	Early Cretaceous	Deinonychus had a big, sickle-shaped claw for attacking prey.
Diplodocus (dih-PLOD-uh-kus)	double beam	plants	26 m (88 ft)	Late Jurrassic	Diplodocus had a hole near the top of its skull whichmay have helped it breathe while feeding.
Hadrosaurus (HAD-ro-SAWR-us)	big lizard	plants	9 m (30 ft)	Late Cretaceous	Hadrosaurus was the first dinosaur to be discovered and recorded in North America.
Megalosaurus (MEG-ah-lo-SAWR-us)	great lizard	meat	9 m (30 ft)	Jurrassic	Megalosaurus was the first dinosaur to be described.
Nodosaurus (no-doe-SAWR-us)	toothless lizard	plants	5.4 m (18 ft)	Late Cretaceous	Nodosaurus had a suit of armor as did its relative, Ankylosaurus.
Ornitholestes (or-nith-o-MY-mus)	bird robber	meat	1.8 m (6 ft)	Late Jurrassic	Ornitholestes was an agile quick-moving dinosaur.
Pachycephalosaurus (pak-ee-SEF-uh-lo-SAWR-us)	thick-headed lizard	plants	4.5 m (15 ft)	Late Cretaceous	Pachycephalosaurus was a bone-headed dinosaur. Rival males may have had butting contests for females

Observe DINOSAURS

1. Draw your dinosaur.

Front view

Side View

2. Make a word list.

3. Write 2 sentences.

My dinosaur is_____ .

My dinosaur has_____ .

Classify Dinosaurs

Sort the dinosaurs into 2 groups.

Group 1

Group 2

All these dinosaurs _____ .

All these dinosaurs _____ .

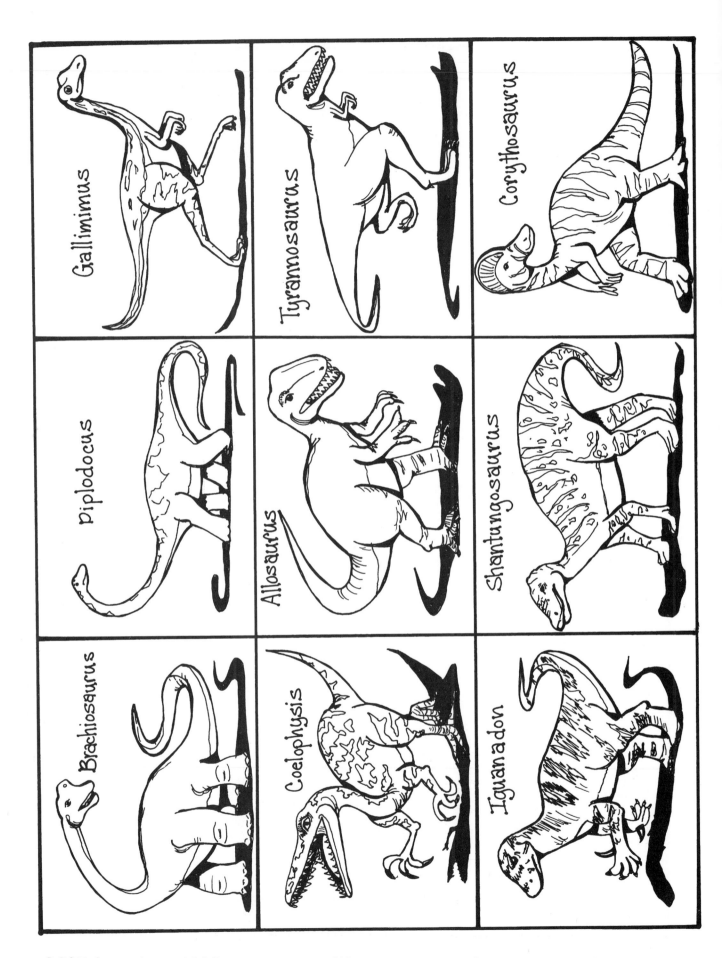

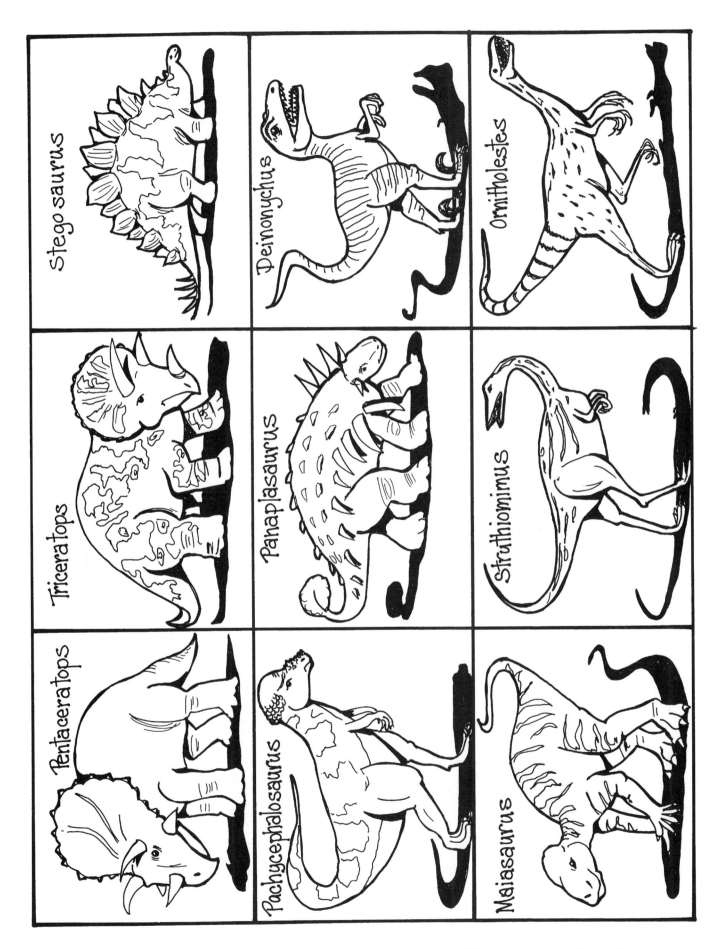

Stegosaurus

Deinonychus

Ornitholestes

Triceratops

Panaplasaurus

Struthiomimus

Pentaceratops

Pachycephalosaurus

Maiasaurus

Communicate about DINOSAURS

1. Look at the dinosaurs. How are they different?
2. Cut out the cards. 20
3. Think how to group them.
4. Glue cards to make a graph.
5. Name each group.

Predict about Fossils

How did dinosaurs become fossils?

Make a prediction. Cut apart the cards. Tape the cards together to tell the story.

Discovered

Dies

Uncovered

Alive

Decays and Sinks

Covered-Turns to stone

Measure

1. Make a paper outline of yourself.

2. Use your outline to measure.

Dinosaur	Long	Tall	Tail

3. Think of another way to measure. Use your hands ✋ a ruler ▭ box 🗃 or cubes ▢▢.

Dinosaur	Long	Tall	Tail

Inferring about Fossils

1. Cut out the bones. ✂
2. Divide the bones fairly.
3. Take turns putting the bones on paper.
4. Work together to make a creature.
5. Glue the bones down.
6. Draw a picture of the creature in its environment.

Name of Creature:

How did this creature move? _____

What did this creature eat? _____

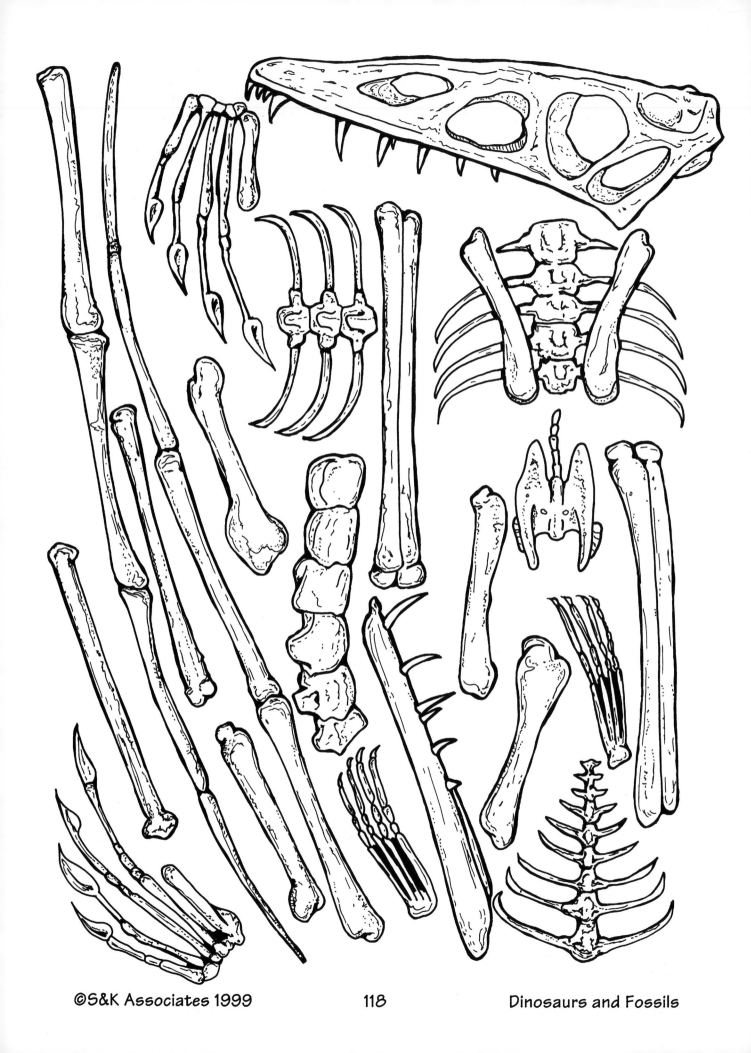

Make a Model of DINOSAURS

1. You have studied fossils and dinosaurs. Now make a model for a mini-museum.

2. Materials List:

3. Draw your model.

Mini-Museum Card

Dinosaur: _____

Size: _____

Food: _____

Protection: _____

Habitat: _____

4. How is your model like a dinosaur?

5. How is your model different from a dinosaur?

DINOSAURS

Define Operationally
What is a dinosaur?

1. Study the 2 dinosaurs. Tell about them.

Stegosaurus has

Deinonychus has

2. Compare the dinosaurs.

Both dinosaurs have

3. What is a dinosaur?

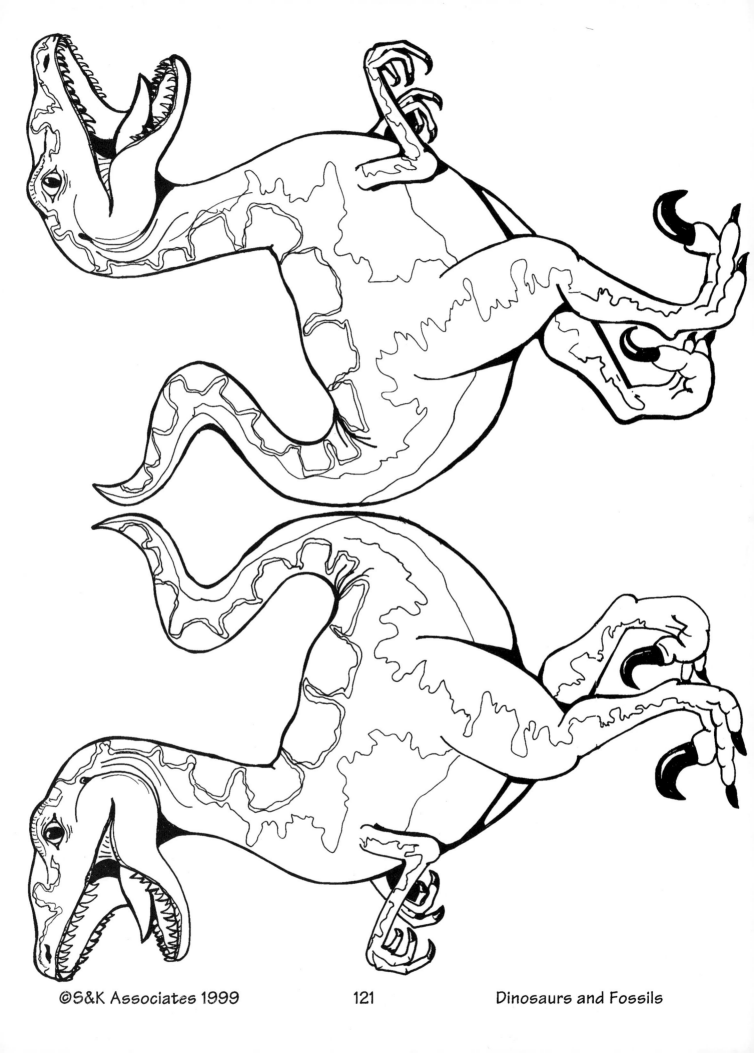

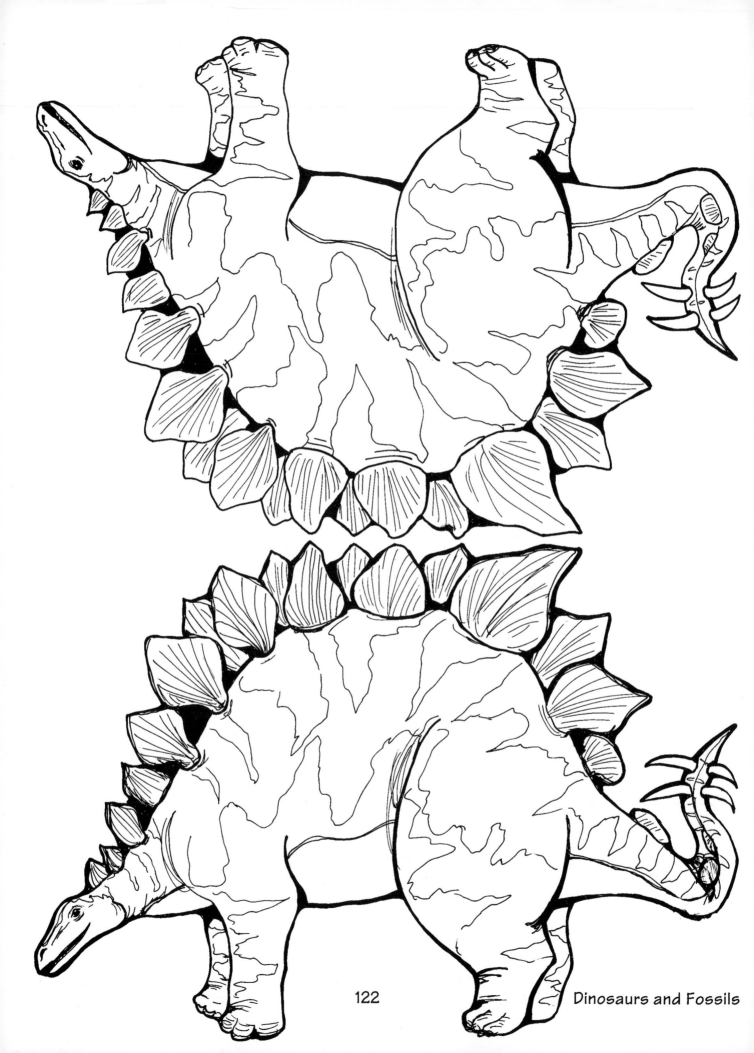

122

Dinosaurs and Fossils

Investigate DINOSAURS

What fossils can we find at this Dinosaur Dig?

1. What do you think you will find?_____

_____

2. Work as a team. Dig carefully!
Draw what you find in each grid.

A.	B.	C.
D.	E.	F.

3. Name what you found in each grid.

A.	B.	C.
D.	E.	F.

Investigate Dinosaurs - Dinosaur Dig

Job Badges — "Cut out the cards and wear as a Job Badge!"

Digger
Use a spoon to carefully dig and uncover fossils.

Cleaner
Use a brush to carefully clean the fossils.

Curator
Place the fossils carefully in your museum.

4. List the steps your team took to dig up the fossils. cut 2

Dinosaur Dig Journal

Name: _____

Date: _____

Science Assessment System

Traditional Assessments	
Type	Advantages
True-False Multiple Choice Matching Items	• Allow many students to be assessed in a short time over a wide range of content areas • Do not require extensive preparation of materials • Can be scored quickly by hand or machine
Completion Short Answer Items	• Answer is constructed rather than chosen lowering the chance of guessing • Misconceptions & partial understanding more easily identified because students do not react to limited number of options • Allow for partial credit
Essay	• Demonstrate students' conceptual understanding & ability to organize and communicate information • Reveal student misconceptions • Lower chance of guessing • Allow for partial credit

Alternative Assessments	
Type	Advantages
Performance Tasks: • Explore • Investigate • Experiment Activities • Performance-Based Activities *Note: Scoring Rubrics & Self-Assessment Checklists are provided for all performance tasks*	• Performance tasks are focused, student centered, & authentic • Students demonstrate the acquisition of inquiry/process skills • Students can generate multiple solutions to problems • Students can connect their knowledge to real world • Assessment format is natural & unobtrusive (closest to instruction & realistic) • Correspond to how problems are commonly encountered and addressed in the real world involving students in research as scientists • Allow for determining if students understand the relevant content and are proficient with inquiry/process skills & can apply skills & concepts to new situations • Rubrics can be presented to students so they understand expectations & can be used to provide feedback to students

Portfolios
 	
• Investigative
  Entry
 	
• Research
  Entry
 	
• Applications
  Entry
 	
• Open Choice
  Entry
	• Activities & products are suggested for inclusion
	• Involve students in assessment of their own achievement & growth by selecting entries & updating their portfolio
	• Evaluation is on the broad picture of each child's growth in science content & process over time
	• Investigative entries require students to design, carry out, and report results of a student initiated question or problem
	
	• Research entries require students to conduct research to investigate & solve a problem of personal or societal significance & make a decision
	• Applications entries require students to use scientific information for a purpose other than research such as an expressive or inventive entry
	• Allow student to submit any piece of classroom work that provides further evidence that is described in the student's reflective summary

Assessments Designed for Multiple Intelligences/ Learning Styles	• Accommodate multiple intelligences with activities appropriate for different learning styles Linguistic Logical-Mathematical Spatial Bodily-Kinesthetic Musical Interpersonal Intrapersonal

Generic Scoring Rubric

4 points   correct, complete, detailed
3 points   partially correct, complete, detailed
2 points   partially correct, partially complete, lacks some detail
1 point   incorrect or incomplete, needs assistance

	4	3	2	1

Total Points _____

Science as Inquiry
The National Science Education Standards

"From the earliest grades, students should experience science in a form that engages them in active construction of ideas and explanations and enhances their opportunities to develop the abilities of doing science."

As a result of activities in grades K-4, all students should develop
* Abilities necessary to do scientific inquiry
* Understandings about scientific inquiry

Abilities necessary to do scientific inquiry	Understandings about scientific inquiry
Ask a question about objects, organisms, and events in the environment.	Scientific investigations involve asking and answering a question and comparing the answer with what scientists already know about the world.
Plan and conduct a simple investigation.	Scientists use different kinds of investigations depending on the question they are trying to answer (describing objects, events, and organisms; classifying them; and doing a fair test-experimenting).
Employ simple equipment and tools to gather data and extend the senses.	Simple instruments, such as magnifiers, thermometers, and rulers, provide more information that scientists obtain using only their senses.
Use data to construct a reasonable explanation.	Scientists develop explanations using observations (evidence) and what they already know about the world (scientific knowledge). Good explanations are based on evidence from investigations.
Communicate investigations and explanations.	Scientists make the results of their investigations public; they describe the investigations in ways that enable others to repeat the investigation.
Ask a question . . .	Scientists review and ask questions about the results of other scientists' work.

Full scientific inquiry involves asking a simple question, completing an investigation, answering the question, and presenting the results to others.

Use the following model for doing **Inquiry Investigations**. The sections can be cut and arranged on a 12x18 piece of construction paper and folded to stand like a mini-science fair display.

My Question

Student questions may emanate from previous investigations, classroom science activities or questions students ask each other. Students can investigate earth materials, organisms, and properties of common objects. Encourage students to ask questions about everything they observe in their environments.

My Guess

This is the student's hypothesis. Students have experiences, understandings, and ideas to draw on to develop conceptions about the way the world works. Encourage students to come up with possible explanations for their questions by asking questions such as, "Why do you think it looks like that? Why do you think it does that?"

My Materials

Young children are developing skills such as how to observe, measure, cut, connect, switch, turn on and off, pour, hold, tie, and hook. Students learn to use simple instruments such as rulers to measure length, height, and depth of objects and materials; thermometers to measure temperature; watches to measure time, beam balances and spring scales to measure weight and force; magnifiers to observe objects and organisms,; and microscopes to observe the finer details of plants, animals, rocks, and other materials. Students also learn to use computers and calculators for conducting investigations. Guide students by asking questions such as, "What would you need to find out the answer to your question?" Suggest equipment and/or materials that students could use and provide assistance by demonstrating how to use simple instruments.

My Plan

Students can design investigations to try things in order to see what happens. Young children can conduct investigations based on systematic observations. At an early age, children tend to focus on concrete results of tests and begin to develop the idea of a "fair test" in which only one variable at a time is changed. Ask questions to help students plan their investigations such as, "What could you do that would answer your question?"

My Observations

Encourage students to use as many senses as applicable for the investigation (tasting and smelling may not be appropriate). Guide students to make systematic observations and organize their data into concrete charts, tables, and graphs. Concrete data displays can be make by placing the objects in rows or columns on a plastic shower curtain with a grid drawn on it. Rows or columns can be labeled with Post-Its®. Young children can draw pictures to show what happened.

My Results

Student observations are used to formulate explanations. This requires critical thinking as students make judgements about the evidence (their observations) in order to come up with an explanation. Guide students to compare their results with their guess (hypothesis) and determine is it was supported or not supported by their data. Students should be assured that it is alright for their hypothesis (guess) not be be supported by their observations. Tell students that it is just as important to find out that things don't work they way they might have thought they did as it is to find out that things do work the way they thought. In fact, often it is when things don't work the way we think they do that we make the greatest discoveries!

My Answer to the Question

The answer to the student's question may or may not be the same as their guess. This is a statement of what the student found out by doing the investigation.

My New Question

In a good investigation, more questions arise where there was only one question before! Ask students, "What other question(s) would you like to have answered after conducting your investigation? Are you wondering about anything else?"

Inquiry Scoring Rubric

4 points   correct, complete, detailed
3 points   partially correct, complete, detailed
2 points   partially correct, partially complete, lacks some detail
1 point   incorrect or incomplete, needs assistance

Scoring Criteria	4	3	2	1
Question can be answered with a simple investigation.				
Guess (hypothesis) suggests a logical answer to the question.				
Listed or drew pictures of all materials used in the investigation.				
Listed or drew pictures of each step required to carry out the investigation.				
Listed observations using as many of the senses as appropriate for the investigation.				
Described results in words or pictures and compared results to guess.				
Answer for the question was based on actual results.				
Listed a new question that could be investigated based on observations during the investigation.				

Total Points _____

My Question

I think _____

My Guess

My Materials

My Plan

Inquiry Model

My Observations

See 👀 _____
Smell 👃 _____
Taste 👅 _____
Touch ✋ _____
Hear 👂 _____

My Results

My results and guess were:
alike   different

My Answer to the Question

_____

My New Question

_____ ?